Activities Manual

for

Seeing Through Statistics

Third Edition

D0140647

Jessica M. Utts
University of California, Davis

THOMSON

BROOKS/COLE

Australia • Canada • Mexico • Singapore • Spain • United Kingdom • United States

Printed in the United States of America
2 3 4 5 6 7 08 07 06 05

Printer: Malloy Incorporated

ISBN: 0-534-39407-8

For more information about our products, contact us at:
Thomson Learning Academic Resource Center
1-800-423-0563

For permission to use material from this text or product, submit a request online at
http://www.thomsonrights.com.
Any additional questions about permissions can be submitted by email to
thomsonrights@thomson.com.

Thomson Brooks/Cole
10 Davis Drive
Belmont, CA 94002-3098
USA

Asia
Thomson Learning
5 Shenton Way #01-01
UIC Building
Singapore 068808

Australia/New Zealand
Thomson Learning
102 Dodds Street
Southbank, Victoria 3006
Australia

Canada
Nelson
1120 Birchmount Road
Toronto, Ontario M1K 5G4
Canada

Europe/Middle East/
South Africa
Thomson Learning
High Holborn House
50/51 Bedford Row
London WC1R 4LR
United Kingdom

Latin America
Thomson Learning
Seneca, 53
Colonia Polanco
11560 Mexico D.F.
Mexico

Spain/Portugal
Paraninfo
Calle/Magallanes, 25
28015 Madrid, Spain

TABLE OF CONTENTS

PART 4: Making Judgments from Surveys and Experiments

TEAM PROJECTS FOR USE IN CLASS

This section contains 10 projects that are designed to be used in a discussion section accompanying a course from *Seeing Through Statistics.* Each project includes a list of necessary supplies, a list of instructions, and a "Team Recording Sheet" for recording your results.

The projects are designed to keep pace with the book at the rate of about one a week for a one quarter, 10-week course. Here is a list of projects, with relevant chapters of the book:

Project 1: *Studying Studies*
This project can be done after completing Chapter 2.
Project 2: *Surveying Telephone Books*
This project can be done after completing Chapter 4.
Project 3: *Designing an Experiment to Study Caffeine and Pulse Rates*
This project can be done after completing Chapter 5.
Project 4: *Data Collection and Description*
This project can be done after completing Chapter 7.
Project 5: *Predicting Correlation*
This project can be done after completing Chapters 10 and 11.
Project 6: *Impaired Functioning*
This project can be done after completing Chapters 12 and 13.
Project 7: *Probability In The Long Run*
This project can be done after completing Chapter 16.
Project 8: *Let's Make A Deal*
This project can be done after completing Chapter 16.
Project 9: *Animal Eyes Confidence Intervals*
This project can be done after completing Chapter 20.
Project 10: *Interpreting Journal Articles*
This project can be done after completing Chapter 23.

PROJECT 1: STUDYING STUDIES

SUPPLIES FOR THIS PROJECT:
Two or more news stories, including at least one about an observational study and at least one about a randomized experiment. Ideally, there will be enough stories so that each story goes to at least two teams and at most five teams. There should be enough copies so that each team has a few copies of its story. If possible, you should also have the original sources from which the news stories were written, such as journal articles or research reports.

Possible news stories are in the Appendix of the book and their original sources are on the accompanying CD. Good choices for this project include News Stories 1, 3, 6, 8, 10, 11, 14, 15, 18, 19 and 20.

INSTRUCTIONS:
1. Divide the class into teams of 3 or 4 people. Enter the names of your team members on your Project 1 Team Recording Sheet.
2. Each team takes a news story. Each story should go to at least two teams, so that you can compare your results with those of another team reading the same story. Each story can go to as many as five teams. This may be useful for a large class.
3. For your team's news story, address the 11 questions/components on the "Team Recording Sheet." All team members can work together, or you can divide up the work. Spend at least half an hour on this.
4. Reconvene as a class. Discuss each story. Have one of the teams briefly describe the study to start the discussion. Divide up the 11 items on the Team Recording Sheet, so that each team that read that story gets to explain some of them to the class. You can randomly choose which team answers which questions.
5. After discussing each question, ask the other teams with that story if they had a different explanation.

Project 1 Team Recording Sheet

TEAM MEMBERS:

1. _____   3. _____

2. _____   4. _____

Is the research based on an observational study or a randomized experiment? How do you know?

What are the explanatory and response variables? Can "cause and effect" be concluded?

If it's an observational study, give an example of a possible confounding variable. If it's a randomized experiment, specify the treatments.

Component 1: The *source* of the research and of the *funding*.

Component 2: The *researchers* who had *contact* with the participants.

Component 3: The *individuals* or objects studied and how they were *selected*.

Component 4: The exact nature of the *measurements* made or *questions* asked.

Component 5: The *setting* in which the measurements were taken.

Component 6: *Differences* in the groups being compared, in *addition* to the factor of interest.

Component 7: The *extent* or *size* of any claimed effects or differences.

Summary: Are the results of the study convincing to you? If not, why not? If so, would you change anything in your life because of the study?

PROJECT 2: SURVEYING TELEPHONE BOOKS

SUPPLIES FOR THIS PROJECT:
- A telephone directory for each team.
- A method for generating random numbers for each team. Many calculators will do this. Or you can put numbered slips of paper in a bag, in which case each team will need paper, scissors, a pen and a bag.
- An overhead transparency and pen for each team.

INSTRUCTIONS:
1. Divide the class into teams of 3 or 4 people. Enter the names of your team members on your Project 2 Team Recording Sheet.
2. Each team chooses a proportion they are going to estimate using the white pages of the phone book. More than one team can choose the same thing, and that would allow you to compare answers. Some possibilities include the proportion who do not list an address, the proportion of last names ending in son (Johnson, Anderson, etc.), the proportion of phone numbers ending in 0 (zero), or the proportion of last names that have only one household listed with that last name. Spend about 5 minutes deciding this. Write your choice on your Team Recording Sheet. Reconvene the class, and have each team report what they are planning to investigate.
3. As a class, discuss various sampling methods from Chapter 4, and how they could be applied to the phone book. For instance, a cluster sample could be done by randomly selecting 10 pages, randomly selecting a column on each of them, then using every listing in that column.
4. Now break into your teams again. Each team must design and execute a sampling plan that will work well to estimate its proportion, including a decision about sample size based on margin of error. As mentioned under "Supplies" your team should have a source of

random numbers, because you will need to use random sampling of some sort.

5. Prepare an overhead transparency explaining the sampling plan and how it was used.

6. After all teams have completed their sampling work and estimated the appropriate proportion, reconvene the class. Have each team presents its sampling plan and results.

7. If multiple teams used the same question, compare their sampling methods and answers. Would you expect each team to get the exact same answer?

8. Discuss whether the methods used were appropriate for the question in each case. For example, if cluster sampling was used to estimate the proportion of names ending in "son" that would not be appropriate, because all of the "Johnsons" are listed together.

Project 2 Team Recording Sheet

TEAM MEMBERS:

1. _____   3. _____

2. _____   4. _____

What proportion are you planning to estimate?

What sampling method are you planning to use? Explain briefly how you will implement the plan.

What sample size are you planning to use and why?

Keep track of your results here, and compute your final proportion.

PROJECT 3: DESIGNING AN EXPERIMENT TO STUDY CAFFEINE AND PULSE RATES

(Adapted from a project suggested by Josh Tabor)

SUPPLIES FOR THIS PROJECT:
- Caffeinated and decaffeinated soda, enough to give half of the class a 4-ounce drink of one and half a 4-ounce drink of the other
- Small cups with two different designs, to hold at least 4 ounces
- A watch with a second hand
- A coin to flip for each pair of participants

INSTRUCTIONS:
The goal of the project is to design and carry out a study to see if soda with caffeine elevates pulse rates more than soda without caffeine.

1. Choose someone to prepare the drinks while the rest of the class discusses the design of the study. The supplies should include two different types of cups, so they can be distinguished. Without allowing the class to see which type of soda is going into which type of cup, the drink preparer chooses one of the two types of cups for the caffeinated soda and one for the non-caffeinated soda. (It's even better if this is done randomly. But make sure all of one type of soda goes into Design 1 and all of the other type goes into Design 2.) Prepare enough cups so that half of the class can have each type. For instance, if there are 30 students in the class, prepare 15 cups of caffeinated soda and 15 cups of non-caffeinated soda.

2. The experiment will use matched pairs. The goal is to have one person in each pair drink caffeine and the other drink a non-caffeinated soda, then to compare their pulse rates after about 10 minutes. As a class, discuss how the pairs will be created. Some possible variables you can use include gender, initial pulse rate, caffeine consumption in the past few hours, etc.

3. After the pairs have been created, flip a coin in each pair to decide who will drink from cup design 1 and who will drink from cup design 2. At this point, you should not know which design contains the caffeine!

4. Everyone should drink the soda at the same time.

5. Discuss other design issues for about 10 minutes so the caffeine has time to take effect. You can do this as a class, or create teams of two matched pairs. The Team Project Sheet has room to address some of these issues. Here are some issues to discuss:
   - How should pulse be measured? Start counting at 0 or at 1? Neck or wrist? Measure for 15 seconds, or 30, or 60?
   - Why was blocking (matched pairs) used? What known sources of variability between people were controlled, that would not have been controlled if we had randomly assigned half of the class to drink each type of soda without using matched pairs? What additional sources of variability might still exist, that you were not able to match on?
   - Why was randomization used to decide who got which cup design rather than allowing you to choose which half of the pair would drink the caffeine? What additional variables, not controlled for by blocking, would be approximately equalized by randomization?
   - Is blinding used in this study? Discuss how it is being used.
   - Would it be better to use the *change* in pulse rate from before drinking the soda to 10 minutes after drinking it, or to just use the pulse rate after the 10 minutes?

6. About 10 minutes after drinking the soda, measure your pulse rates. Within each pair, note whether the person with cup Design 1 or cup Design 2 had the higher pulse rate. For how many pairs did Design 1 result in a higher pulse? For how many pairs did Design 2 result in a higher pulse?

7. The drink preparer should now reveal which design held the caffeinated soda.

Project 3 Team Recording Sheet

TEAM MEMBERS:

1. _____ 3. _____

2. _____ 4. _____

You may want to work as a team during the 10 minutes the caffeine is taking affect. Discuss and answer these questions.

Why is it important to determine rules for how to measure pulse rate within each matched pair?

What known sources of variability were controlled by using matched pairs?

What known sources of variability were you not able to match on?

What unknown sources of variability are hopefully evened out somewhat by using randomization? (One example could be food consumed recently.)

Was the experiment double or single blind? Could it be double blind?

PROJECT 4: DATA COLLECTION AND DESCRIPTION

(Adapted from a suggestion from Janet Rutherford.)

SUPPLIES FOR THIS PROJECT:

Use the letter *T* to be number of teams, and *N* to be number in class.

- *T* paper bags
- *T* pieces of paper, each with a different question on it (suggestions are in Step 1 of the instructions.)
- *NT* small slips of paper, divided into *T* sets of *N*, each to be used for recording one data value
- Measuring instruments, such as a ruler or tape measure
- *T* overhead transparencies and pens
- *T* or more regular pens or pencils

INSTRUCTIONS:

The purpose of this project is to illustrate natural variability, different shapes for measurement variables, and ways of summarizing variables. To do this, you will measure several variables on each person in the class, and have each team summarize the results of one variable.

1. To prepare for this project, *T* "data collection stations" should be set up around the room. At each station, in a prominent place that can be read, place a piece of paper with the question that will be asked of each visitor to that station. Also place a paper bag, *N* slips of paper, one for each visitor, and a few pens or pencils. If needed to answer the question at that station, place a measuring instrument, such as a ruler or tape measure. Here are some possible questions; the class can generate more as needed:
   - What month were you born? Record the number, with January = 1, etc.
   - What is your mother's height in inches?
   - How many compact discs (CDs) do you own?

11

- How many (natural) teeth do you have?
- What is your hand span (in *centimeters*) for your dominant hand? Your dominant hand is the one you normally use to write. Your hand span is the length of your fully stretched hand from the tip of your thumb to the tip of your little finger. Also record if you are male or female.
- How many brothers and sisters do you have (total, including step and half siblings)?
- How many minutes does it take you to get from your residence to school (or to your first class of the day)?

2. Divide into teams. Each team should start at a different station. Each person should use a slip of paper to record the answer to the question at that station and place the slip in the paper bag provided. To help assure anonymity, use the pens or pencils provided rather than your favorite purple or aqua ink pen!

3. Rotate around to all of the stations, with one team visiting each station at a time.

4. After everyone has visited all of the stations, convene as a class. Discuss what shape you think each of the variables will have, and whether you think there will be major outliers.

5. Reconvene the teams. Each team gets one of the paper bags with the N slips of paper in it, and a transparency and pen. For your set of measurements, follow the instructions on your team sheet, and put the results on your overhead transparency:

6. Each team will present the summary to the class using the overhead projector. Discuss the interesting features of your variable.

7. Discuss all of the variables measured. Which had the most natural variability? Which do you think was most prone to measurement error? Were any of them hard to summarize? Do you think responses were biased for any of them?

Project 4 Team Recording Sheet

TEAM MEMBERS:

1. _____   3. _____

2. _____   4. _____

Using the slips of paper for the variable you are summarizing, do the following here and on your transparency:

Create a stemplot. Make sure you decide how many numbers to put on the stem based on the range of the data. Make sure each number on the stem represents the same size interval.

Describe the shape. Is it close to the shape the class expected that measurement to have?

Create a five number summary

Identify any unusual features including outliers, patterns, or whatever else you see

PROJECT 5: PREDICTING CORRELATION

SUPPLIES FOR THIS PROJECT:
A transparency and pen for each team, with axes for a scatterplot
T slips of paper *for each person* in the class (T = number of teams)

INSTRUCTIONS:
1. Divide into teams of 3 or 4 people. Record your team members' names on the Team Recording Sheet.
2. The teacher will give each team a type of scatter plot to aim to create; see Step 3 for further explanation. You may be assigned to create a scatter plot with moderate positive correlation and no outliers, with strong negative correlation and outliers, with near zero correlation, and so on.
3. Once you have been assigned your target type of scatter plot, your job is to think of two variables for which you think the class responses will result in that type of scatter plot. Here are some examples of the kinds of x and y variables you might ask about:
   x = typical ounces of water you consume per day
   y = weight
   x = your mother's age when you were born
   y = your age (to nearest month if possible) when you went out on your first date
   x = number of hours you studied last weekend
   y = GPA
4. Each person should be given T slips of paper. Each team should ask their two questions of the class. Each class member should write their answers to those two questions and give them to the team that asked. Make sure you specify which value is for x and which is for y.
5. Draw a scatterplot of your results on the team sheet, and on the transparency. Display your scatter plot and discuss whether it fits the pattern you were assigned and any interesting results you see.

Project 5 Team Recording Sheet

TEAM MEMBERS:

1. _____   3. _____

2. _____   4. _____

What type of scatter plot were you assigned to create?

What two variables did you ask people?

Draw your scatter plot here.

Were you successful in creating the type of scatter plot assigned?

PROJECT 6: IMPAIRED FUNCTIONING

SUPPLIES FOR THIS PROJECT:
- A 16 ounce plastic cup and a few wadded pieces of paper for each team
- A calculator for each team

NOTE: Teams for this project should have 3 people.

PURPOSE:

Being impaired from alcohol consumption, lack of sleep, stress and so on can affect performance and increase the risk of making mistakes. The purpose of this exercise is to simulate a situation in which being impaired increases the risk of poor functioning. You will conduct an experiment, calculate risk under normal and impaired conditions, find the relative risk under the two conditions and determine if there is a statistically significant relationship between impairment (or not) and making a mistake, for you.

INSTRUCTIONS:
1. Determine your *dominant eye*. Do this by holding your arm out-stretched in front of you and making a circle with your thumb and index finger. Look at a distant spot through the circle. Now close one eye, then the other eye. You should <u>still be able to see the spot</u> when your <u>dominant eye</u> is open, but the spot will have <u>shifted out of the circle</u> when your <u>non-dominant eye</u> is open.
2. Divide into teams of 3 people. Your team will need a cup and a few wadded pieces of paper.
3. Decide who will do each of three tasks (you will then switch roles). One task is to hold the cup on the floor. Another task is to record the results. The third task is to try to drop the wadded paper into the cup from a standing position.
4. Each team member should do a total of 40 tries to drop the paper

into the cup, 20 in an <u>unimpaired</u> condition and 20 in an <u>impaired</u> condition, *alternating* between the two conditions. Flip a coin to decide which condition to do first. Here are the two conditions:
<u>Unimpaired</u>: Stand above the cup and attempt to drop the paper into the cup with your dominant hand and both eyes open.
<u>Impaired</u>: Stand above the cup and close your *dominant* eye. Attempt to drop the paper into the cup with your *non-dominant* hand.

5. The recorder should keep track of your performance and record your results on your team sheet.

6. When everyone has finished, for each person compute the *chi-square statistic* (see the shortcut formula below the table on the team sheet), the *risk of missing the cup* under *each* condition, and the *relative risk* of missing the cup for the two conditions. Put the *baseline risk* (unimpaired condition) in the denominator.

7. As a class, draw a picture of the relative risk values. (A stem and leaf plot may work well, or a histogram.) Discuss the comparative merits of doing separate relative risks for everyone, versus combining the results for everyone into one overall risk for each condition and one relative risk. Also, determine how many people (what proportion) had a statistically significant relationship.

Project 6 Team Recording Sheet

TEAM MEMBERS:

1. _____   3. _____

2. _____

Fill in the number of times the target was hit and missed under each condition in the following table. Compute the chi-square value as shown. This is a "shortcut formula" that works for a table with 2 rows and 2 columns. Compute the risk of missing under each condition, then find the relative risk. Repeat for each team member. (Use a separate sheet.)

Condition:	Hit Target	Missed Target	Total
Unimpaired	$A =$	$B =$	20
Impaired	$C =$	$D =$	20
Total	$E=A+C=$	$F=B+D=$	40

$$\text{Chi-square value} = \frac{40(A \times D - B \times C)^2}{(20 \times 20)(E \times F)} = \frac{(A \times D - B \times C)^2}{10(E \times F)} =$$

Your results are statistically significant if the chi-square value is greater than 3.84.

Risk when Unimpaired = B/20 = _____ (this is your baseline risk)

Risk when Impaired = D/20 = _____

Relative Risk = (Risk when Impaired)/(Risk when Unimpaired) = _____

PROJECT 7: PROBABILITY IN THE LONG RUN

SUPPLIES FOR THIS PROJECT:
If there are *N* participants in the class you will need:
- *N/2 dice*
- *N/2 pennies*
- *2 calculators*

INSTRUCTIONS:
As a class, discuss the idea of probability as long run relative frequency. The purpose of this project is to watch this happen.

1. Choose 6 volunteers to be the *data accumulators* for the class. They will go to the board. Their jobs are: penny recorder, penny calculator, penny grapher; die recorder, die calculator and die grapher. Give a calculator to each of the two people with "calculator" jobs.
2. The two "recorders" should each create three columns on the board, as follows:

Number of Successes (S)	Number of tries (T)	Relative frequency of success = S/T
Leave room for one entry per class member in each column.		

3. The two graphers should create the axes for a plot on which they will keep a running graph of the relative frequency of success. Label the vertical axis "relative frequency" and the horizontal axis "number of tries."
For the penny graph:
   - The scale on the vertical axis ranges from about 0.3 to 0.7
   - Draw a flat line starting at 0.5 on the vertical axis and extending to the right. (This represents the true probability of a head.)
   - Label the horizontal axis to start at 0 and end at 20 times the number of people in the class, with marks at 20, 40, etc.

For the graph for the dice:
- The scale on the vertical axis ranges from about 0 to 0.4
- Draw a flat line starting at 1/6 (.1667) on the vertical axis and extending to the right. (This represents the true probability of rolling a 6.)
- Label the horizontal axis to start at 0 and end at 20 times the number of people in the class, with marks at 20, 40, etc.

4. Divide into pairs. Each pair gets one penny and one die. One person in each pair should toss the penny 20 times, then roll the die 20 times, while the other person keeps track of the results. For the penny, a "head" counts as a success. For the die, rolling a "6" counts as a success. Record the number of successes in each endeavor for the 20 tries. Complete Step 5, and then switch roles and repeat.

5. After each person completes 20 tries of each task, they should go to the board and report their results to the data accumulators. As each team reports, the "recorder" should update the first two columns (number of successes and number of tries), the calculator should update the 3rd column (relative frequency of successes) and the grapher should update the graph showing the *accumulated* proportion of successes for results of all the participants so far.

6. Each person in each pair should complete one round of tossing pennies and rolling dice. After reporting both of your results to the recorders, you will accumulate results of your own two sessions by finding the individual relative frequencies of success for the two sets of 20 tries, then the combined relative frequency for the 40 tries of each task for your pair. Use your Team Recording Sheet.

7. As a class, review the results on the board. Did the relative frequency of success for each of the two tasks converge to the true probabilities of 0.5 and 1/6? Did they get close? Discuss how the results for the whole class should be more accurate overall than the results for the 40 tries for each pair.

Project 7 Team Recording Sheet

TEAM MEMBERS:

1. _____   2. _____

Team Member 1's Results:

Number of successes (heads) in 20 coin tosses: _____

Relative frequency of success for the coin tosses: _____

Number of successes (rolling a 6) for the 20 roles of the die: _____

Relative frequency of success for the roles of the die: _____

Team Member 2's Results:

Number of successes (heads) in 20 coin tosses: _____

Relative frequency of success for the coin tosses: _____

Number of successes (rolling a 6) for the 20 roles of the die: _____

Relative frequency of success for the roles of the die: _____

Combined Results:
Number of successes (heads) in 40 coin tosses: _____
Relative frequency of success for the coin tosses: _____
Number of successes (rolling a 6) for the 40 roles of the die: _____
Relative frequency of success for the roles of the die: _____

PROJECT 8: LET'S MAKE A DEAL

SUPPLIES FOR THIS PROJECT:

There are *T* teams.

- *3T paper or plastic cups that cannot be seen through, labeled #1, #2 and #3 for each team*
- *2T pennies*
- *T dice*

INSTRUCTIONS:

This is the "Monty Hall" problem, based on the game show "Let's Make a Deal" with host Monty Hall. The goal of this project is to simulate the game, illustrating the optimal strategy for winning.

1. As a class, discuss how the game worked on TV. You are a contestant on a game show. There are three doors. Behind one of the doors is a new car and behind each of the other two doors is a goat. You pick a door, and you will win the prize behind that door. However, you get a second chance to guess. Say you pick door #1. The host of the show (Monty Hall) now tells you that behind door #3 is a goat. You are given the choice of remaining with door #1, or switching your choice to door #2. Should you change your choice to door #2? Does it matter? (Don't discuss the answer yet!)

2. Divide into teams of 3 or 4 people. Each team should have three plastic cups numbered one through three and two pennies (one to hide and one to use for randomization). Choose one team member to start out as Monty Hall. The team member to the right of Monty Hall will be the first contestant and will play 10 rounds. The team member to the right of that person will record the results, using the table on the Team Recording Sheet. See the next step for how to play a round. After the first player finishes 10 rounds, he or she becomes Monty Hall and the next player does 10 rounds. Continue until all team members have had a turn.

3. Here is how the game is played.
   - The contestant turns his/her back to Monty Hall and completely closes his/her eyes. All other team members should close their eyes as well, so as not to give subtle cues to the contestant!
   - Monty Hall rolls a die and places the penny under cup #1 if the roll is a 1 or 2, under cup #2 if the roll is a 3 or 4, and under cup #3 if the roll is a 5 or 6. (The cups are inverted on the table.)
   - Monty Hall then flips a coin and remembers the result for later (to randomly choose which cup to reveal). The die and coin must be removed from the table before anyone else sees them.
   - The contestant turns around and chooses the cup that he/she believes hides the penny. If the contestant chooses the correct cup, then both of the other cups are empty. In this case, Monty Hall should reveal the lower numbered cup if his/her earlier coin flip resulted in heads and the higher numbered cup otherwise.
   - If the contestant chooses an empty cup, then only one of the other cups is empty. In this case, Monty Hall should reveal the only empty cup (but without letting anyone know it's the only empty cup).
   - The contestant then decides whether to change his/her choice.
   - Monty Hall reveals the penny. The recorder must keep track of whether or not the choice was changed *and* whether or not the contestant won on that round.
4. When all contestants have played, compile the results for the team into the table on the Team Recording Sheet. Each team should spend a few minutes discussing whether they think it is more beneficial to switch cups, to keep the original choice, or it doesn't matter.
5. Reconvene the class and accumulate results into a table similar to the one the Team Recording Sheet. Determine the proportion of wins when the contestants switched cups and when they did not switch cups. Which choice was made more often?

6. Stop reading right now if you haven't played yet!! The best strategy for winning will be revealed, which is to switch cups. Here is an easy way to explain the answer. Without loss of generality, assume that the contestant starts by choosing cup #1. There are three equally likely possibilities:

Penny under:	Probability	Contestant wins by:
Cup #1	1/3	Not Switching
Cup #2	1/3	Switching
Cup #3	1/3	Switching

Therefore, the contestant will win 2/3 of the time by switching and only 1/3 or the time by not switching.

Project 8 Team Recording Sheet

TEAM MEMBERS:

1. _____   3. _____

2. _____   4. _____

Fill in the results for each person, then the table for the team.

Name: _____   Name: _____

	Switched	Didn't
Won		
Lost		
Total		

	Switched	Didn't
Won		
Lost		
Total		

Name: _____   Name: _____

	Switched	Didn't
Won		
Lost		
Total		

	Switched	Didn't
Won		
Lost		
Total		

TEAM STATISTICS:

	Switched	Didn't	Total
Won			
Lost			
Total			
Proportion Won = Won/ *Total*			

PROJECT 9: ANIMAL EYES CONFIDENCE INTERVALS

SUPPLIES FOR THIS PROJECT:
- Enough "animal eyes" or substitute so that each pair of participants can have five. The goal is to have something that can land face up or face down, but where the probability of landing face up is not already known. A good example is "animal eyes," which are used to put eyes on stuffed animals and can be found in a craft store. The eyes can land with the pupil face up or face down.
- A calculator for each team
- A grid for displaying confidence intervals, which is a longer version of something like this, on a transparency, and two colors of pens:

.20   .25   .30   .35   .40   .45 **.50**   .55   .60   .65   .70   .75   .80

.20   .25   .30   .35   .40   .45 **.50**   .55   .60   .65   .70   .75   .80
95% confidence intervals for probability of pupil landing face up

When used, each confidence interval is drawn horizontally across one row of the table. Use one pen color for individual confidence intervals and a different color for team confidence intervals. You need enough rows for each individual *and* each team to draw a horizontal confidence interval.

PURPOSE:
The purpose of this project is to estimate an unknown probability. If the animal eyes are used, we are estimating the probability that the eye will land with the pupil facing up when the eye is tossed. A secondary purpose is to notice how the width of a confidence interval

depends on the number of observations. A third purpose is to illustrate how multiple confidence intervals all estimating the same probability are not all exactly the same.

INSTRUCTIONS:

1. Divide into teams of 4, then divide into pairs within the team. Each pair needs 5 animal eyes. (To flip 20 times, for a total of 100 flips.)
2. One partner goes first. Flip the eyes repeatedly until you have 100 flips. The other partner records the number of times they land "eyes up." Enter the data in the table on the Team Recording Sheet.
3. Carry out the calculations as instructed in the table, resulting in a 95% confidence interval for the true probability that the eye will land with the pupil up.
4. Each person should draw his/her confidence interval on the grid on the overhead using the same pen color.
5. Calculate the confidence interval for the team, as shown in the table.
6. Draw the team confidence interval on the overhead using a new pen color, with the same color used for all team confidence intervals.
7. Calculate the confidence interval for the entire class.
8. Discuss what has been demonstrated with this project. Which confidence intervals are the widest – the ones for each individual, the teams, or the whole class? Why? Are all of the confidence intervals the same? Are they centered on the same place? Are they all of the same width, for each person? For each team?
9. Based on all of the results, what do you think the true probability is for the eyes (or substitute) landing face up?
10. Do you think it was appropriate to combine the results for everyone, or should they have been kept separate? Explain.

Project 9 Team Recording Sheet

TEAM MEMBERS:

1. _____ 3. _____

2. _____ 4. _____

1. Toss the eyes repeatedly until you have done 100 tosses. (You can toss them 5 at a time, 20 times.) Your partner records the number of times they land "eyes up." Enter your data in the table below.
2. Carry out the calculations as instructed in the table below, resulting in a 95% confidence interval for the probability of landing eyes up.
3. Draw your confidence interval on the grid on the overhead using the pen color you are instructed to use for the individual intervals.
4. Calculate the confidence interval for the team, as shown in the last row of the table.
5. Draw the team confidence interval on the overhead.

Team Member:	X = # with "Eye Up" (out of 100 tosses)	Proportion with eye up $= \hat{p} = \dfrac{X}{100}$	$S.D.=\sqrt{\dfrac{\hat{p}(1-\hat{p})}{100}}$	95% C.I. $\hat{p} \pm 2(S.D.)$
1				
2				
3				
4				
Team Results	Number with eye up= _____ Total tosses $n =$ _____	$\hat{p} =$ proportion with "eye up" =	$S.D.=\sqrt{\dfrac{\hat{p}(1-\hat{p})}{n}}$ =	95% C.I. $\hat{p} \pm 2(S.D.)$

28

PROJECT 10: INTERPRETING JOURNAL ARTICLES

SUPPLIES FOR THIS PROJECT:
Abstracts from journal articles and/or news stories about studies, one for each team. Two examples are provided after the Team Recording Sheet. If there is time to prepare in advance, class members could bring in examples for use in this project. You could also use some of the Case Studies in the Appendix and on the CD accompanying the book.

INSTRUCTIONS:
1. Divide into teams of 4 to 6 people. Each team is to be given an article. The same article can be given to more than one team, but there should be at least 3 articles used.
2. Your team should discuss the six questions shown on the Team Recording Sheet for your study. Allow about 20 minutes for discussion. Each question should be assigned to one team member to present the answer to the class. Record brief answers to the questions on your Team Recording Sheet.
3. Reconvene the class for discussion. Each team will answer the six questions for the study they were assigned. If multiple teams were assigned the same study, take turns having them answer each of the questions.

Project 10 Team Recording Sheet

TEAM MEMBERS:

1. _____   4. _____

2. _____   5. _____

3. _____   6. _____

Assign each team member to explain one of the following issues to the class for the report you are discussing. Discuss them all as a team for about 20 minutes. If there are only five team members, combine questions 1 and 2. If there are only four team members, combine questions 1 and 2 and questions 3 and 4.

Question 1: What type of study was done? Include a discussion of all of the following if they are relevant: type of sample, use of randomization, use of blocking or pairing, case-control, retrospective or prospective. Include any weaknesses you see in the design.

Question 2: Discuss possible confounding variables, possible interacting variables, and the extent to which they were considered by the authors. How do you think they might affect the results?

Question 3: Discuss what hypothesis tests were done or implied. Specify the null and alternative hypotheses, the results in terms of p-value, and the results in terms of statistically significant findings. (For instance, you might say that a statistically significant relationship was found between treatment and outcome, with p-value = .003.) *If no hypothesis tests were done, you can use a confidence interval presentation to formulate a hypothesis test. Pick a different one from the person who does question 4.*

Question 4: Discuss what confidence intervals were presented. Include an explanation of what they represent, and how they are related to any conclusions made from the study. *If no confidence intervals were given, discuss another hypothesis test, different from the one in Question 3.*

Question 5: Explain what Type 1 and Type 2 errors would be for the main question of interest in this study, and what their consequences would be. Which type of error would be more serious? (Even if confidence intervals only are presented, there is an implicit hypothesis test.)

Question 6: Give an overall description of what was learned from the study, and what cautions should accompany the results.

Here are two samples of the type of material to use for this project:

Project 10, Example 1: The material below is quoted directly from the summary at the beginning of the article.

Survival in Academy Award Winning Actors and Actresses

Donald A. Redelmeier, MD; and Sheldon M. Singh, BSc

Annals of Internal Medicine, 15 May 2001, Volume 134, No. 10, 955-962

 Background: Social status is an important predictor of poor health. Most studies of this issue have focused on the lower echelons of society.

Objective: To determine whether the increase in status from winning an academy award is associated with long-term mortality among actors and actresses.

Design: Retrospective cohort analysis.

Setting: Academy of Motion Picture Arts and Sciences.

Participants: All actors and actresses ever nominated for an academy award in a leading or a supporting role were identified (n = 762). For each, another cast member of the same sex who was in the same film and was born in the same era was identified (n = 887).

Measurements: Life expectancy and all-cause mortality rates.

Results: All 1649 performers were analyzed; the median duration of follow-up time from birth was 66 years, and 772 deaths occurred (primarily from ischemic heart disease and malignant disease). Life expectancy was 3.9 years longer for Academy Award winners than for other, less recognized performers (79.7 vs. 75.8 years; P = 0.003). This difference was equal to a 28% relative reduction in death rates (95% CI, 10% to 42%). Adjustment for birth year, sex, and ethnicity yielded similar results, as did adjustments for birth country, possible name change, age at release of first film, and total films in career. Additional wins were associated with a 22% relative reduction in death rates (CI, 5% to 35%), whereas additional films and additional nominations were not associated with a significant reduction in death rates.

Conclusion: The association of high status with increased longevity that prevails in the public also extends to celebrities, contributes to a large survival advantage, and is partially explained by factors related to success.

Project 10, Example 2: This is quoted directly from the Reuters Health website. Tuesday May 29, 2001, 1:41 PM ET

Hopelessness Linked to Higher Risk of Mortality

NEW YORK (Reuters Health) - A lack of hope for the future is associated with a higher death rate among older men and women, a team of researchers report. "Our findings confirm and extend the results of previous studies suggesting that, in a US sample comprising older women and men of Mexican and European origin, hopelessness is associated with an increased risk of...mortality," according to Dr. Stephen L. Stern, a psychiatrist at the University of Texas Health Science Center in San Antonio, and colleagues.

The study, which was funded by the National Institutes of Health, was published in the journal *Psychosomatic Medicine*.

The researchers sought to explore whether feelings of hopefulness act as a key to survival and whether those who lack hope are at higher risk of death. From 1992 to 1996, Stern's team asked 795 men and women aged 64 to 79 in San Antonio how they felt about the future and whether or not they were ``hopeful about the future.'' About half of the study participants were European-Americans and half were first, second or third generation Mexican-Americans.

Around 9% of the respondents answered no, they had no hope about the future, while the remaining 91% said that they did have hope about the future.

Mexican-American men were twice as likely to report feeling hopeless as European-American men, with 12% of the Mexican-American men saying they were hopeless compared with only 6% of the European-American men. Men and women were equally likely to report hopelessness.

When tracked an average of 5 years later by reviewing death certificates in 1999, the researchers found that 29% of the people who said they felt hopeless had subsequently died, compared to only 11% of those who said they were hopeful. None of the study participants had committed suicide, but had primarily died from either cancer or from heart disease.

The investigators measured whether depression among people who felt hopeless was responsible for their higher death rate. After adjusting for other risk factors, they found that depression was not associated with mortality. But people who suffered from both depression and hopelessness may be at the greatest risk of mortality, the report indicates.

The researchers theorized that hopeless people might have higher death rates because they may suffer from biochemical abnormalities, such as decreased immune function or abnormal platelet function.

The study was limited in that it only measured hopelessness with a simple yes or no answer, the authors note. It also did not examine the reasons why people said they were hopeless.

Stern and colleagues conclude that because this question may provide relevant information about a patient's risk of death, doctors should consider asking it during a checkup. '"Are you hopeful about the future?' might be a useful screening question to include in the evaluation of older patients," they write.

Original Source: *Psychosomatic Medicine* 2001;63:344-351.

ACTIVITIES USING THE APPLETS ON THE CD

This section contains descriptions and activities for the six applets on the CD accompanying the book. The applets are from CyberStats, an online Statistics course. They are designed to be used interactively, so make sure you turn on your computer and start playing!

Here are the applets, and the relevant sections of the book for each one:

The Sampling Applet
The most relevant material is in Section 4.4, but it also relates to material in Chapter 3, the remainder of Chapter 4, Chapter 7, Chapter 16 and Chapter 19.

The Empirical Applet
The relevant material is in Chapter 8.

The Correlation Applet
The relevant material is in Chapters 10 and 11.

The Sample Means Applet
The relevant material is in Section 19.3.

The TV Means Applet
The relevant material is in Section 19.3.

The Confidence Level Applet
The relevant material is in Chapters 20 and 21.

The Sampling Applet

Relevant text material is in Section 4.4, but additional related material is in Chapter 3, the remainder of Chapter 4, Chapter 7, Chapter 16 and Chapter 19. The activities below refer to these other chapters where relevant.

What the Sampling Applet Does

The purpose of the applet is to illustrate taking simple random samples. The applet displays 100 stick figures representing a population of 55 females (blue figures) and 45 male (red figures). They are of different heights, with a mean height of 68.0 inches. When you click the **Sample** button, a simple random sample of 10 individuals is selected. If you set the speed as "Slow" or "Fast" you can watch the individuals being selected. If you set the speed as "Batch" the entire sample will appear at once.

Notice that after taking each sample, the 10 individuals are shown on the right. The mean height for the *sample* and the percent of females in the *sample* are displayed below the sample. When you have chosen a sample you can see for yourself if each of the individuals is male or female, along with the individual heights.

If you click on "Display Results" you will see the accumulated list of mean heights and percent females for all of the samples you have taken so far. Do not click on "Start Over" unless you want to eliminate all of those prior results. However, the individual sets of 10 heights are not stored, so if you want those you need to record them.

Playing With the Sampling Applet

1. Choose one of the stick figures to represent you.
    a. Explain why your chance of making it into the sample is 1 in 10 for each sample of 10 individuals selected.
    b. If you take 20 samples, how many times do you think you will make it into the sample? Will it always be the same number of times? In other words, if you take another 20 samples, will you be in the same number of them as you were for the first 20 samples?
    c. Take 20 samples. How many times were you selected? When you learn about probability in Chapter 16 you will learn that in the long run, you should be selected in about 1 of 10 samples, but in the short run (your 20 samples) you may be selected more or less often. It wouldn't be surprising to find that you were selected 0, 1, 2, 3 or even 4 or 5 times.

2. Take 20 samples, and then click on "Show Results." You should see a popup window with 20 lines, where each line gives the mean height and percent female for one sample.
    a. Describe how the "mean height" changes. For instance, what are the lowest and highest values? How do the mean heights from the sample compare to the mean height for the population of all 100 individuals, which is 68.0 inches? Based on a single sample of 10 individuals, would you be likely to get a good estimate of the mean height for the population?
    b. Describe how the "percent female" changes. For instance, what are the lowest and highest values? How do the values compare to the population percent of females, which is 55%? Based on a single sample of 10 individuals, would you be likely to get a good estimate of the percent of females in the population?

c. For samples of size 10, the size used in this applet, is it possible to have a sample mean height of 68.0 (the population mean height)? Is it possible to have a sample percent female of 55% (the population percent female)? Explain.

d. Does the mean height for the sample appear to be related to the percent of females in the sample? Would you expect them to be related? Explain.

3. See pages 46-47 of the book for the definitions of types of variables.

   a. In this applet, two variables are measured for each individual selected. What are they?

   b. For each of the two variables you named in part a, explain which of these terms apply to it: categorical, nominal, ordinal, measurement, interval, ratio, discrete, continuous.

4. See the definitions of the following terms on page 61 of the book, and then explain what they are for this applet.

   a. A unit.

   b. The population.

   c. A sample measurement.

   d. The sampling frame.

5. Read the explanation of how to take a systematic sample, on page 67 of the book.

   a. How would you use the stick figures to take a systematic sample of five individuals from this population?

   b. Take a systematic sample of five individuals (by hand). Explain what you did in enough detail that someone else could find your sample.

   c. How many males and how many females were in your sample from part b?

d. In using a sample to estimate the mean population height and percent female, would results from a systematic sample be biased? Explain.

6. Take a sample and notice the heights of the shortest and tallest individuals in the sample. Continue taking many samples until you are fairly certain you know what the heights are for the shortest and tallest individuals in the population. Now take one more sample of 10 individuals. Based on a single sample, would you be likely to learn what the heights are for the shortest and tallest individuals in the population?

The following activities can be performed after you have covered Chapter 7 in the book.

7. Use the applet to generate 20 simple random samples of size 10. Make sure you don't click the "Start Over" button. Click the "Display Results" button. (If you have access to Minitab or other software you can cut and paste the results into the program and use if for the following.)
   a. Make a stemplot of the sample means. What shape is it?
   b. Find the mean and the median of the sample means. What are they? Are they close to the population mean of 68.0 inches? Are they close to each other?
   c. Make a tally of the number of samples with each possible percent of females (30%, 40%, etc). What is the mode value? Is this surprising or is it what you would expect?

Understanding What Happened with the Sampling Applet

Playing with this applet should give you some feeling for what can be expected from simple random samples. Here are specific details about each of the activities above.

Notes about Activity #1 for the Sampling Applet: In general, your chance of getting into a simple random sample is the proportion found by dividing the size of the sample by the size of the population, which in this case is 10/100 or 1/10. For most real surveys the population is much larger than 10 times the size of the sample, so if you have never been chosen for a survey it isn't surprising. Out of 20 samples, you would get into 2 of them on average, but it wouldn't be surprising if you weren't in any of the samples, nor would it be surprising if you were in as many as 5 of them. You will learn more about how to quantify probabilities in Chapter 16.

Notes about Activity #2 for the Sampling Applet: The mean height should vary from sample to sample, but not nearly as much as the heights vary within each sample. Very short and very tall people should cancel each other out in each sample. It is possible to get a sample mean of 68.0 inches, and in most cases the sample mean won't be off by more than about an inch and a half. So the sample mean is a pretty good estimate for the population mean. The sample percents of females on the other hand, are more variable. You should see them ranging from about 30% to 80%. It's impossible to get the actual population 55% because the sample percent can only be a multiple of 10. In answer to part d, yes, the two should be related. The higher the percent of females, the smaller the average height should be. That's because the females in the population are in general shorter than the males.

Notes about Activity #3 for the Sampling Applet: The two variables are height and gender. Height is a measurement, interval, continuous variable. Gender is a categorical, nominal variable.

Notes about Activity #4 for the Sampling Applet: A unit is one stick figure. The population of units is the collection of 100 stick figures, and the population of measurements includes their genders and heights. The sample measurements are the heights and genders of the 10 individuals selected. The sampling frame is the collection of 100 stick figures.

Notes about Activity #5 for the Sampling Applet: For a systematic sample of size 5, choose one individual from each set of 20. Choose a random starting point in the first two rows and then sample every 20th individual after that. This would be equivalent to choosing the individual in the same position in each set of two rows. There shouldn't be any bias because the stick figures are not arranged in any pattern.

Notes about Activity #6 for the Sampling Applet: The shortest and tallest heights are 63 inches and 75 inches. In any sample of 10 individuals, you may not see one or both of these heights even though they both occur more than once in the population.

Notes about Activity #7 for the Sampling Applet: The stemplot should be approximately bell-shaped. The mean and median should be similar to each other and should be close to 68 inches. You will learn more about what to expect of a collection of sample means in Chapter 19 in the book, which is called "The Diversity of Samples from the Same Population." The percents of females should range from 30% to 80%, with 50% and 60% as the most likely values. You may occasionally see a sample with 20% or 90% females, but you would rarely see one with 0%, 10% or 100%. About one in 240 samples would have 10% females.

The Empirical Applet
Relevant text material is in Section 8.4

What the Empirical Applet Does

The purpose of the applet is to demonstrate how well the Empirical Rule works for different shaped data sets. The applet includes histograms for eight variables, using data collected from students at the University of California at Davis and Penn State University. When you highlight a variable in the box on the left, the applet displays a histogram with vertical lines drawn at one standard deviation on either side of the mean (pink lines) and two standard deviations on either side of the mean (blue lines). The sample mean and standard deviation are provided. The numerical intervals "Mean ± s.d." and "Mean ± 2s.d." are provided (with the letter "s" used to indicate the standard deviation), and the percent of the sample data falling into each interval is shown.

The variables are:

Sleep = Hours of sleep the night prior to the survey.

TV Hours = Hours spent watching TV in a typical week.

Dad's height = Father's height in inches.

Exercise = Hours spent exercising in a typical week.

Ideal height (females) = Females response to asking their ideal height.

Alcohol = Number of alcoholic beverages consumed in a typical week.

Handspan = Width of stretched hand from thumb tip to baby finger tip, measured in centimeters. Shown separately for females and males.

Playing With the Empirical Applet

1. Refer to the Empirical Rule on page 154 of the book.
    a. Fill in the blanks: Approximately _____% of the values fall within 1 standard deviation of the mean in either direction and _____% of the values fall within 2 standard deviations of the mean in either direction, *if* the population has a _____ shape.
    b. Now refer to the applet. Approximately what percent of the data should fall into the pink interval (Mean ± s.d.) if the Empirical Rule holds? Approximately what percent of the data should fall into the blue interval (Mean ± 2s.d.)?

2. For each of the eight variables, characterize the shape of the distribution as approximately bell-shaped, somewhat skewed, or extremely skewed, and note whether there are major outliers, minor outliers or no outliers. Look carefully to make sure you notice any outliers that are at the extremes of the histograms.

	Shape?	Outliers?
a. Sleep	_____	_____
b. TV Hours	_____	_____
c. Dad's Height	_____	_____
d. Exercise	_____	_____
e. Ideal Height	_____	_____
f. Alcohol	_____	_____
g. Handspan (females)	_____	_____
h. Handspan (males)	_____	_____

3. If there are major outliers in a data set:
    a. Will they cause the standard deviation to be larger or smaller than it would be for the data without the outlier(s)? Explain.

b. Will the widths of the intervals used in the Empirical Rule (Mean ± one s.d., etc.) be bigger or smaller than they would be without the outlier(s)? (This answer follows directly from your answer in part a.)

c. Using your answer in part b, do you think the percent of the data covered by the intervals Mean ± one s.d., etc, will be higher or lower when outliers are present than it would be otherwise? Explain.

4. Fill in the values found using the applet, for the percent of the data in each of the following intervals.

	Mean ± one s.d.	Mean ± two s.d.
a. Sleep		
b. TV Hours		
c. Dad's Height		
d. Exercise		
e. Ideal Height		
f. Alcohol		
g. Handspan (females)		
h. Handspan (males)		

5. Study the results from Activities 2 and 4.
   a. What can you conclude about how well the Empirical Rule works for these data sets? In what situation (shape, outlier status) does is work best?
   b. In what situation (shape, outlier status) does it work least well? Using your answer to Activity 3, explain why this is the case.

6. For which of the two intervals, Mean ± one s.d. or Mean ± two s.d., did the Empirical Rule work less well when outliers and a skewed shape were present?

Understanding What Happened with the Empirical Applet

The Empirical Rule says that approximately 68% of a bell-shaped population (or normal curve) should fall within one standard deviation of the mean, and about 95% should fall with two standard deviations of the mean. Outliers will cause the standard deviation to be larger than it would be otherwise, thus making the intervals wider. This means that they are likely to cover a larger percent of the data than they would without the extreme outlier(s), especially for the central part of the data, the interval Mean ± one s.d..

For this applet, the data sets and Empirical Rule results can be characterized as follows:

- The data sets that are approximately bell-shaped with no major outliers are *Sleep, Dad's height (minor outlier), Ideal Height* and *Handspan (males)*. For these, the Empirical Rule works fairly well.
- The *Handspan (females)* data set is clumped around the upper values, with a few minor outliers making it look somewhat skewed to the left. The standard deviation is slightly inflated because of the outliers, so the intervals are a bit wider than they would be otherwise, and include more of the data set than the Rule predicts.
- *TV Hours, Exercise* and *Alcohol* are highly skewed to the right with extreme outliers. As a consequence, the standard deviation is large, creating wide intervals that cover much more of the data than the Empirical Rule predicts for bell-shaped data. This isn't surprising, because these data sets are not even close to being bell-shaped.

In summary, notice that the Empirical Rule works well in the situation for which it is intended, namely, bell-shaped data, but does not work well for highly skewed data sets.

The Correlation Applet

Relevant text material is in Sections 10.3 and 11.1

What the Correlation Applet Does

The purpose of the applet is to enable you to explore how the value of
the correlation changes when you add points to a plot. There are three
scatter plot frames (with no points) presented with the applet. Each
one has a "target" correlation that you are supposed to try to achieve.
The plots start out without any points on them and your goal is to add
points to the plot until the value of the correlation is within .05 of the
target correlation specified for that plot. You need to place at least 15
points on the plot to qualify as having achieved the target. The target
correlations include one positive (.5), one negative (–.8) and zero.

The default when you start the applet is "Add points." Place the mouse
cursor where you want to add a point, and then click. The point should
appear on the scatter plot. If you hold the mouse button down you will
see the actual coefficients for the point. As soon as you have placed
two points on the scatter plot, a correlation value will appear. (The
correlation for two points is always 1.0.) As you continue to add points,
the correlation value is updated. If you want to remove a point, click
the radio button for "Delete points." Then click near the point you want
to delete. To completely start over, click "Clear."

The message "Goal Reached!" will appear when you have successfully
created a correlation value to within .05 of the target correlation with
at least 15 points on the plot.

Playing With the Correlation Applet

1. Use the first scatter plot frame in the applet to do the following.
   a. Create a set of points that has a correlation value within 0.05 of the target correlation of 0.5, using at least 15 points. Do this *without* including any extreme outliers.
   b. Create a set of points such that the correlation is about 0.8 for the first 14 points. Then add a single outlier that lowers the correlation to about 0.5. You may have to play with adding and deleting points until you figure out how to do this. Characterize the type of outlier that made this happen. Was it in line with the other points?
   c. Create a set of points such that the correlation is close to 0 for the first 14 points. Then, add a single outlier that increases the correlation to about 0.5. You may have to play with adding and deleting points until you figure out how to do this. Characterize the type of outlier that made this happen. Was it in line with the other points?
   d. Using the results from parts b and c, characterize the affect different types of outliers have on the value of the correlation. Explain what type of outlier inflates correlation and what type of outlier deflates correlation.

2. Use the second scatter plot frame in the applet to do the following.
   a. Create a set of points that has a correlation value within 0.05 of the target correlation of –0.8, as instructed, using at least 15 points. Do this *without* including any extreme outliers.
   b. Create a set of points by putting a tight cluster of points in the upper left corner for the first 14 points, such that the correlation is no more than about 0.2 in absolute value (i.e. it's between –0.2 and +0.2). Then add a single outlier such that the

correlation increases to be within 0.05 of the target of –0.8, just by adding the outlier. Characterize what happened.

c. Create a scatter plot illustrating two groups where the correlation is positive within each group, but the correlation for the combined groups is within 0.05 of the target correlation of –0.8.

d. Based on the results of parts b and c, describe two ways to create a strong negative correlation that are both misleading, if the interpretation of the strong negative value is that as one variable increases steadily the other decreases steadily.

3. Using the third scatter plot frame in the applet, play with various ways in which you can place points to get within 0.05 of the target correlation of 0. Describe some ways to do this. For instance, would the best straight line through the points always be flat and have a slope of zero, or are there other ways to get close to 0 as the correlation value?

4. Using the third scatter plot frame in the applet, create a scatter plot for which there is a clear curved pattern to the points, but for which the correlation is within 0.05 of the target of 0.

5. In statistics, an "influential point" is one that has substantial influence on the statistical results, in this case, on the value of the correlation. Look at the two scatter plots accompanying Thought Question 1 of Chapter 11 of the book (p. 200), each containing an outlier. Reproduce them approximately on one of the scatter plot frames. What affect does each type of outlier have on the correlation? Does one type appear to be more influential than the other? Use this to characterize what "influential points" might look like in the context of correlation.

6. Using any of the scatter plot frames, place points that follow:
   a. An almost exact U-shaped pattern. What is the correlation?
   b. An X-shaped pattern. What is the correlation?
   c. What type of pattern does correlation measure? Based on your results in parts a and b, do you think it's possible to have a strong relationship between two variables, but still have a correlation near 0? Explain.

7. Using any of the scatter plot frames, place 10 points on the scatter plot and note the correlation. Now place another 10 points, almost on top of the original 10 points. Did the correlation change much? Do you think correlation is affected by the number of points included?

8. This activity is based on a famous example that illustrates what can go wrong if you simply look at a correlation without seeing the scatter plot that produced it. (Anscombe, F.J., 1973, "Graphs in statistical analysis, *American Statistician,* p. 17-21.) Plot the following four sets of 11 points, approximately, and notice the patterns of the points and the correlations. Discuss what happened. (Plots 1, 2 and 3 all use the same values for x.)

Use same X's for plots 1,2,3	Y's for plot #1	Y's for plot #2	Y's for plot #3	X's for plot #4	Y's for plot #4
20	33	31	24	40	23
25	47	47	27	40	26
30	62	61	31	40	28
35	38	73	34	40	36
40	60	81	38	40	39
45	78	88	41	40	40
50	70	91	45	40	47
55	73	93	48	40	49
60	98	91	52	40	55
65	66	87	97	40	58
70	90	81	58	95	95

Understanding What Happened with the Correlation Applet

Activities 1, 2, and 5 illustrate the impact of outliers on correlation, and the effect of combining groups inappropriately. Activity #1 illustrates that a single outlier can substantially increase or decrease the correlation. In part b, you should have points close to a straight line with positive slope, then add an outlier that is not at all close to the line to decrease the correlation. In part c, you should start with a set of points in the lower left or upper right region of the plot that show very little pattern. Then add an outlier in the opposite corner of the plot from the cloud of points, to substantially increase the correlation.

The second activity illustrates two ways that you can be misled with correlations. The one in part b is similar to that illustrated in the first activity, in which one point can substantially inflate the correlation. The example in part c shows how combining groups and looking at the combined correlation can be very misleading, compared with looking at the correlation separately within each group. Activity 5 continues to emphasize how outliers can increase or decrease the correlation. An influential point that increases correlation is one that you can visualize as "pulling" the line towards it. One that decreases correlation is a single exception to the linear pattern of the other points.

Activities 3, 4 and 6 illustrate that correlation measures a linear (straight line) relationship only. A scatter plot in which the points closely follow another pattern, such as a U-shape or X-pattern, but for which the best straight line through the points is flat, will have a correlation value near 0.

Activity 7 illustrates that correlation is not affected by number of points. Activity 8 shows 4 very different plots, all with correlation near 0.82. Correlation is only appropriate for Plot #1.

The Sample Means Applet
Relevant text material is in Section 19.3

What the Sample Means Applet Does

The purpose of this applet is to illustrate the Rule for Sample Means.
Remember that there are two situations for which the Rule for Sample
Means applies – see page 362 of the book. The first situation is
illustrated by this applet – the population of measurements is bell-
shaped and a random sample of any size is measured. (The second
situation is illustrated by the next applet, "TV Means.")

The Rule for Sample Means (p. 363 of the book) provides the following
information. If we take numerous samples of the same size (call it n for
number in sample) and find the sample means for each of them, then:
- The frequency curve, which can be represented by a histogram of all
  the sample means, should be approximately bell-shaped.
- The mean of the frequency curve should have the same mean as the
  original population.
- The standard deviation of the frequency curve should be the
  original population standard deviation divided by the square root of
  the sample size n.

The applet presents a population (in pink) with mean = 8.0 and standard
deviation = 5.0, like the weight loss example in Section 19.3. It takes
simple random samples from this population. For each sample, it
computes the sample mean and then creates a histogram using all of the
sample means so far (in red). You make these choices:
- The sample size, n, for each sample. (Default is n = 25 observations.)
- The number of samples you want to take: 1, 10, 100 or 500.
- The speed of sampling. "Batch" means the histogram is created at
  the end, after all of the samples have been selected.

Playing With the Sample Means Applet

Note: The scale for the histogram is set from 4 to 11, so the applet cannot display all of the results for samples smaller than 15 or so, and the histogram will not have enough bars of interest for samples larger than about 100.

1. Begin using the applet with the default sample size of 25. Generate 500 samples.
   a. What does the Rule for Sample Means predict that the mean and standard deviation of the resulting histogram will be?
      Mean:_____   Standard deviation: _____
   b. Remember from the Empirical Rule that almost all observations should fall within 3 standard deviations of the mean. If the mean and standard deviation you predicted in part a are correct, what are the lowest and highest values you should see in the histogram? In other words, what range spans from 3 standard deviations below the mean to 3 standard deviations above the mean?
      (Mean − 3 s.d. = _____) to (Mean + 3 s.d. = _____)
   c. Does the mean of the histogram (in red) for your 500 samples look like it is what you predicted in part a?
   d. Does the histogram span the range you predicted in part b?
   e. Does the histogram have the shape it is expected to have? Explain.
   f. Suppose the number of pounds of weight lost on a particular diet, for the population of people who follow the diet, is approximately normal with mean of 8 pounds and standard deviation of 5 pounds, as in this applet. Use the histogram you generated in this activity to answer the following. If one simple random sample of 25 people is selected, is it likely that the average weight loss for them would be as low as 4 pounds? Is it likely that the average weight loss for them would be as high as 9 pounds? Explain.

2. Change the "# Observations per sample" to 100. Generate 500 samples.
   a. What does the Rule for Sample Means predict that the mean and standard deviation of the resulting histogram will be?
   Mean:_____   Standard deviation: _____
   b. If the mean and standard deviation you predicted in part a are correct, what are the lowest and highest values you should see in the histogram? In other words, what range spans from 3 standard deviations below the mean to 3 standard deviations above the mean? (Mean – 3 s.d. = _____) to (Mean + 3 s.d. = _____)
   c. Does the mean of the histogram (in red) for your 500 samples look like it is what you predicted in part a?
   d. Does the histogram span the range you predicted in part b?
   e. Does the histogram have the shape it is expected to have? Explain.
   f. Compare the results of this activity to the results for Activity #1. What is the same and what is different?

3. Play with different sample sizes (#Observations per sample), generating 500 samples each time. What stays about the same in the histogram as the sample size changes, and what does not? Does the shape stay about the same? Does the mean stay about the same? Does the standard deviation stay about the same? Do your observed results concur with what the Rule for Sample Means would predict?

4. Return to the default sample size of 25.
   a. Generate 10 samples, then click on the "Clear" button and generate another 10 samples. Do this several times. Does the histogram look the same for each set of 10 samples?
   b. Repeat part a with 100 samples each time instead of 10. Don't forget to click on the "Clear" button between sets of samples.

c. Repeat part a with 500 samples each time instead of 10. Don't forget to click on the "Clear" button between sets of samples.

d. Now generate several histograms of 1000 sample means. Do this by generating 500 samples, then another 500 samples without hitting the "Clear" button between them. Then clear the results and generate another two sets of 500 for a total of 1000 samples. Do this several times.

e. About how many samples did it take before the shape of the histogram became clear? Was 10 enough? Was 100 enough?

f. About how many samples did it take before the histogram barely changed between sets of samples? Was 100 enough? Was 500 enough? Was 1000 enough?

5. Generate 2000 samples of 25 observations by clicking on the "500" button four times without clearing the results between clicks. The resulting histogram should approximate the smooth frequency curve that the Rule for Sample Means predicts for this situation. Find the figure in the book that illustrates in theory what the frequency curve should look like for this situation. Compare it to your histogram. Are they similar? Are you convinced by this that the Rule for Sample Means works in this situation? Explain.

6. Repeat Activity 5, but generate only 100 samples. Is the resulting histogram similar to the theoretical frequency curve figure shown in the book? Should it be? Explain.

7. Suppose you owned a franchise of the weight loss clinic described in Section 19.3. The regional manager stops by and uses your records to select a simple random sample of 25 of your clients. He finds that the mean weight loss for the 10-week program for this sample was only 7 pounds. He says that isn't good enough because it should be 8 pounds. Using this applet, how could you defend your results?

Understanding What Happened with the Sample Means Applet

Notes about Activity #1 for the Sample Means Applet: The Rule for Sample Means predicts that the histogram will be bell-shaped with a mean of 8 and standard deviation of 1.0, so it should range from about 5 to 11. One random sample of 25 is unlikely to have a mean as low as 4, but likely to have a mean as high as 9.

Notes about Activity #2 for the Sample Means Applet: The Rule for Sample Means predicts that the histogram will be bell-shaped with a mean of 8 and standard deviation of 0.5, so the histogram should range from about 6.5 to 9.5. The shape and mean should be the same as for Activity #1, but the histogram should cover only about half of the range.

Notes about Activity #3 for the Sample Means Applet: The Rule for Sample Means predicts that the histogram will remain approximately bell-shaped, and you should have observed that. The mean, or center of the histogram, is predicted to stay at 8, and you should have observed that. However, the standard deviation is predicted to change as the number of observations changes, and you should observe that. Remember that the standard deviation is predicted to be the population standard deviation, which is 5, divided by the square root of the number of observations. Thus, the standard deviation should be about 1.0 for $n = 25$ and about 0.5 for $n = 100$. You can observe this change by noting how spread out the histogram is.

Notes about Activity #4 for the Sample Means Applet: The approximate bell-shape of the histogram probably won't be clear for many of the sets of 10 samples. The shape should be more clear for sets of 100 samples, but it should still be somewhat choppy. The shape should definitely be clear with 500 sets of samples, but even then, the shape will change somewhat from set to set. With 1000 sets of samples, the picture should be changing just slightly between sets.

Notes about Activity #5 for the Sample Means Applet: Figure 19.5 on page 364 of the book illustrates the theoretical frequency curve. If you generate 2000 samples the histogram should be a good approximation for the theoretical curve. For instance, it should have a symmetric bell-shape, should be centered on 8, and should range from 5 to 11.

Notes about Activity #6 for the Sample Means Applet: With 100 samples, the histogram probably won't look like the theoretical frequency curve. There aren't enough samples to cover all of the possibilities and the histogram will look choppy.

Notes about Activity #7 for the Sample Means Applet:
You could show the regional manager the applet, taking 1 sample at a time of 25 observations. Show him how the sample mean is different each time, and that it isn't at all uncommon to find a sample mean of 7 pounds or less, even though the population mean is 8 pounds.

The TV Means Applet
Relevant text material is in Section 19.3

What the TV Means Applet Does

The purpose of this applet is to illustrate the Rule for Sample Means in the second situation described on page 362 of the book:
"The population of measurements of interest is not bell-shaped, but a large random sample is measured. A sample of size 30 is usually considered 'large,' but if there are extreme outliers, it is better to have a larger sample."

The applet can be used to illustrate that the Rule for Sample Means does *not* work well in this situation unless a large sample is used. Read the description of the results given by the Rule for Sample Means in the section for the previous applet, "What the Sample Means Applet Does." Thus, the histogram of sample means from thousands of *large samples* from a non-normal population should be bell-shaped. The mean should be the original population mean, and the standard deviation should be the original standard deviation divided by the square root of the number of observations, *n*. The shape should not hold for *small* samples.

The TV Means applet presents a population of responses to the question in a student survey, "How many hours of television do you watch in a typical week?" The population mean = 8.352 hours and the population standard deviation = 7.723 hours. The histogram of responses (in pink) shows that they are highly skewed to the right.

The applet takes simple random samples from this population. The choices (sample size, number of samples and speed) are the same as for the previous applet.

Playing With the TV Means Applet

1. Generate 2000 samples with $n = 4$ observations. (Use 4 batches of 500 samples without clearing.) Observe the histogram (in red).
   a. Should the histogram be approximately bell-shaped? Explain.
   b. Is the histogram approximately bell-shaped? If not, is the shape what you would expect? Explain.
   c. Although the Rule for Sample Means does not address this, the mean and standard deviation for the resulting frequency curve should be as described in the Rule for Sample Means even for small samples. In this case, what should the mean and standard deviation of your resulting histogram be? Mean:_____ s.d._____
   d. Check to see if the mean of your histogram is close to what you predicted in part c. (If you think of the bars of the histogram as being filled with liquid, and the horizontal number line as a see-saw, the histogram would balance at the mean. This may help you visualize where the mean falls.)
   e. Would you expect the Empirical Rule to work in this situation, thus allowing you to predict that the histogram should be covered by the interval Mean ± 3s.d.? Explain why or why not.

2. Repeat Activity #1 using $n = 16$.

3. Repeat Activity #1 using $n = 25$.

4. Repeat Activity #1 using $n = 36$.

5. Repeat Activity #1 using $n = 49$.

6. Based on the results of these activities, about how large should n be for the Rule for Sample Means to work in this situation?

Understanding What Happened with the TV Means Applet

No matter what sample size you use, the mean for the histogram should be about the same as the population mean, 8.352 hours. The standard deviation should be about equal to the population standard deviation of 7.723, divided by the square root of the sample size. Thus, the standard deviation for the red histogram should be about:

3.86 for $n = 4$	1.29 for $n = 36$
1.93 for $n = 16$	1.10 for $n = 49$
1.54 for $n = 25$	

The shape for $n = 4$ should be skewed to the right, but not as extremely so as for the original population. For $n = 16$ and $n = 25$, the skewed shape should still be evident, but the histogram should be getting closer to bell-shaped. For $n = 36$ and $n = 49$ the bell-shape should be completely evident.

The Empirical Rule should not work well for predicting the interval the histogram should cover unless the histogram is approximately bell-shaped. Thus, it should not work well for $n = 4$ at all. This makes it hard to verify whether or not the expected standard deviation of 3.86 holds, because the applet does not provide the standard deviation for the sample means used to create the histogram. For the larger sample sizes, it would in theory be possible to verify the predicted standard deviations because the Empirical Rule should hold. However, the applet uses a small number of bars for those sample sizes, so unfortunately it may be difficult to verify the results in those cases as well.

The lesson from this applet is that the shape of the frequency curve of sample means depends on the sample size if the population is skewed. You can see that the convention of defining "large" as 30 or so seems to work well for this example.

The Confidence Level Applet

Relevant text material is in Chapters 20 and 21 (Sections 21.1 and 21.4)

What the Confidence Level Applet Does

The purpose of this applet is to illustrate the concept of the *confidence level* associated with a confidence interval, discussed on page 400 of the book. The applet generates random samples of size $n = 64$ from a normally distributed population with a mean of 170 and standard deviation of 20. This is similar to the population of weights of college-age men. It then displays a confidence interval for the population mean based on the generated sample. The true population mean of 170 is fixed in place and indicated by a vertical red line.

You specify the confidence level, let's call it C. The default when you open the applet is $C = 68$, for a 68% confidence interval. Change the confidence level by moving the slider below where the level is displayed.

The formula used for the confidence interval is Mean $\pm z$ (SEM), where:
- Mean = sample mean for the 64 observations, different each time.
- z = the value of the standard normal curve such that C% of the normal curve is between $-z$ and $+z$; such as 1.96 for 95% confidence.
- SEM = standard error of the mean = 20/8 = 2.5. Notice that 20 is the population standard deviation, and 8 is the square root of the sample size of 64.

Each time you click "sample!" a new sample of $n = 64$ is generated, and a confidence interval for the population mean is displayed. A running tally is kept for the number of intervals that capture the population mean of 170. The percent of successful intervals is also displayed. You can start over by clicking "reset!". You can ask the applet to repeatedly generate samples by clicking "animate!". It will continue until you click "stop!".

Playing With the Confidence Level Applet

1. Generate one sample with the Confidence level set at 68%. Now move the slider to increase the confidence level. What happens to the center and the width of the interval? Explain why this happens.

2. Generate one sample at a time with the Confidence level set at 68% until you get an interval that does *not* cover the true mean of 170 (the red line). Now move the slider to increase the confidence level. (The applet uses the same sample to create confidence intervals with the new confidence levels.) Does the interval ever cover the true mean of 170? Explain what has happened.

3. Set the confidence level to 68%.
    a. Generate 5 samples, one click at a time. Do not reset the results each time. What percent of the 5 intervals covered the true mean of 170? Is this what you would expect? What percents are possible based on 5 intervals? Is 68% a possibility?
    b. Click "animate!". Stop the process after about 20 intervals and answer the same questions as in part a, this time for 20 intervals.
    c. Continue until about 100 intervals have been generated and answer the questions in part a, this time for 100 intervals.
    d. What did you discover in parts a to c of this activity?

4. This is a repeat of Activity #3 with the confidence level set at 95%. Set the confidence level to 95%.
    a. Generate 5 samples, one click at a time. Do not reset the results each time. What percent of the 5 intervals covered the true mean of 170? Is this what you would expect? What percents are possible based on 5 intervals? Is 95% a possibility?
    b. Click "animate!". Stop the process after about 20 intervals and answer the same questions as in part a, this time for 20 intervals.

c. Continue until about 100 intervals have been generated and answer the questions in part a, this time for 100 intervals.

d. What did you discover in parts a to c of this activity?

5. Set the confidence level at 99%. Click "animate!". Go get a cup of coffee, fold your laundry, etc, for 10 minutes or so. Come back when the applet has generated at least 1000 intervals. Report the results.

6. Based on what you have learned from these activities:
   a. Explain in your own words what the confidence level is.
   b. If 10 experimenters each take random samples and construct 90% confidence intervals, will 9 of the intervals cover the truth and one interval not cover the truth? Explain.
   c. If 1000 experimenters each take random samples and construct 90% confidence intervals, about what percent will cover the truth?
   d. For any given experiment and confidence interval, will the researcher know whether it has covered the truth?

7. In a real experiment, only one sample is taken and only one confidence interval is constructed. Explain in your own words what the "confidence level" represents for a single confidence interval.

8. Explain why sometimes a low confidence level, such as 68%, might be preferable and sometimes a high confidence level, such as 99% might be preferable. What is the trade-off between the two?

9. If random samples of the same size are taken from a population and a 95% confidence interval for a mean is created each time, which of the following change and which stay the same?
   a. The endpoints of the interval.
   b. The true population mean.

Understanding What Happened with the Confidence Level Applet

Notes about Activity #1 for the Confidence Level Applet:
The center of the interval is at the sample mean, which doesn't change for a given sample. The width of the interval increases as the confidence level increases because the multiplier, *z*, increases.

Notes about Activity #2 for the Confidence Level Applet:
As you move the slider, you should eventually capture the true mean. (The applet allows a maximum of 99% as the confidence level, so in about 1 out of 100 cases, you could get a sample for which even a 99% confidence interval didn't capture the true mean of 170.)

Notes about Activity #3 for the Confidence Level Applet:
With only 5 intervals, it is not possible to have exactly 68% of them capture the true mean, because that would be 3.4 intervals. The only possibilities are 0%, 20%, 40%, 80% or 100%. With 20 intervals, 68% would be 13.6, so that's not exactly possible either. However, it is possible to have 65% or 70% capture the true mean. With 100 samples you should get close to the long-run 68% coverage rate. What you should learn from this is that the confidence level applies in the long run, not the short run.

Notes about Activity #4 for the Confidence Level Applet:
With only 5 intervals, it is not possible to have exactly 95% of them capture the true mean, because that would be 4.75 intervals. The possibilities are 0%, 20%, 40%, 80% or 100%. With 20 intervals it is possible to have 95% capture the true mean, if exactly 19 did and 1 did not. With 100 samples you should get close to the long-run 95% coverage rate. What you should learn from this is that the confidence level applies in the long run, not the short run.

Notes about Activity #5 for the Confidence Level Applet:
The percent of intervals that captured the truth should be very close to 99%.

Notes about Activity #6 for the Confidence Level Applet:
The confidence level represents the proportion or percent of times the procedure works in the long run, over thousands of times it is used. For any set of 10 confidence intervals with 90% confidence, there is no guarantee that exactly 9 of them will be correct (cover the truth). With 1000 intervals, the percent correct should be very close to the confidence level, 90% in this case. For any specific interval, there is no way to know if it is correct or not without knowing the true population value. And if you knew the true population value, you wouldn't need a confidence interval!

Notes about Activity #7 for the Confidence Level Applet:
The confidence level represents your confidence in the procedure. For any given sample and corresponding interval, we won't know if the procedure worked or not.

Notes about Activity #8 for the Confidence Level Applet:
The tradeoff is that the higher the confidence level the wider the interval will be. Therefore, if you want to be almost certain your interval covers the truth you should use a high level of confidence, such as 99%. If you want a rough idea of the magnitude of the true mean, but not complete confidence that you have captured it, a 68% confidence interval may be sufficient.

Notes about Activity #9 for the Confidence Level Applet:
The endpoints of the interval move around but the true mean doesn't.

THE EDUCATED CITIZEN'S GUIDE TO STATISTICS

Statistical methods are extremely useful for finding out about the world around us, as I hope you will learn as you read *Seeing Through Statistics*. As an educated citizen, you should be able to read about statistical studies, polls and events involving uncertainty and make sound conclusions. The entire book is devoted to helping you learn how to do that. This section provides special reinforcement for twelve of the most important concepts, with focus on ones that often are misunderstood or misrepresented in the news. These questions and activities will help you focus on the ideas and test your understanding. The emphasis here is on how these concepts relate to your daily life.

PART 1: Finding Data in Life
Concept 1: Do the results of that study apply to me?
Concept 2: If I change my behavior will it matter?
Concept 3: Can I believe the results of that poll?

PART 2: Finding Life in Data
Concept 4: Can I be unusual and still be normal?
Concept 5: Relationships are not always what they appear to be!
Concept 6: Combining groups: Beware of Simpson's Paradox!

PART 3: Understanding Uncertainty in Life
Concept 7: Wow! Was that coincidence really amazing?
Concept 8: Don't get your probabilities backwards!
Concept 9: Knowing when not to trust your intuition.

PART 4: Making Judgments from Surveys and Experiments
Concept 10: What's important may not be significant!
Concept 11: What's significant may not be important!
Concept 12: Why does one out of 20 foods seem to cause cancer?

Concept 1:
Do the results of that study apply to me?

The Basic Idea

The results of a study can be extended to a larger group if the units in the study are representative of the larger group *for the questions asked in the study*. One way to make sure this is true is to use a random sample from the larger group as the units in the study, and this is done commonly for surveys and polls. However, using a random sample is rarely feasible for randomized experiments, and generally volunteers are recruited and used instead. For observational studies, convenience samples often are used, such as students or hospital patients.

Test Your Understanding

Question 1.1: Suppose a randomized experiment is done using college student volunteers between the ages of 18 and 22 as the participants. Give one example where the results could reasonably be extended to all people in that age group, and one example where the results could not. For each of your examples explain what the explanatory variable (treatments) and the response variable are.

Question 1.2: Find two examples of polls in the news. (Many websites have them, including those for the large polling agencies.) Find one example for which the results apply to a group larger than just those who participated in the poll, and one example for which the results do not. Explain why you reached that conclusion in each case.

Question 1.3: Read the Original Source for Case Study 1 on the CD accompanying the book. The article is called "Alterations in Brain and

Immune Function Produced by Mindfulness Meditation." Do you think the results of the study apply to you? Why or why not?

Question 1.4: Read the news stories and the abstract for the Original Source for Case Study 15 on the CD accompanying the book. The website http://www.ucl.ac.uk/epidemiology/hbu/habit.html explains what "HABITS" is, and that the sample consisted of "approximately 5,000 secondary school teenagers from 36 schools all over South London." Do you think the results apply to children in this age group (11 to 12 years old) in your country and region? Explain.

Suggested Answers

Suggested Answer for Question 1.1: The key is whether the relationship being studied is one for which college student volunteers are likely to differ from others in that age group. Example 1 (could reasonably be extended): Students are randomly assigned to watch a comedy show or an action show for half an hour, then their blood pressure is measured to see if action shows tend to elevate blood pressure more than comedy shows. Example 2 (could not reasonably be extended): Students are randomly assigned to attend either a weekend motivational seminar or a weekend sports seminar. The response measured is the hours spent working (studying) in the next month, and the two groups are compared to see if the motivational seminar led people to work harder. The results could not reasonably be extended to others in this age group who weren't motivated to work hard in general.

Suggested Answer for Question 1.2: In general, if the poll was based on random digit dialing or some other random sampling method, as are the ones done by the large polling agencies, then the results can be extended to a larger group. If the poll used volunteers, as do many "quick polls" on websites, then in most cases the results apply only to those who participated. The key is whether the participants represent a larger group *for the questions asked* and in most cases, in polls using a volunteer sample they do not.

Suggested Answer for Question 1.3: The participants were right-handed, healthy employees of a biotech company in Madison, Wisconsin. There were males and females in the study. If you are right-handed and healthy, the results probably apply to you. If you are left-handed and healthy they probably still apply. (The handedness factor was important in this study so that the same region of the brain could be measured for all participants. If left-handers had participated, the side of the brain measured would probably have to be reversed.) There is probably nothing inherent in the effects of meditation on the brain and the immune system that would be different for the people in this study than it would be for you. One caveat is that they were willing volunteers, and if the idea of meditation is repulsive to you, then it probably wouldn't work!

Suggested Answer for Question 1.4: The results probably apply to children in urban and suburban areas in cultures similar to that in South London, which would include at least most western cultures. The results may not apply to rural children in western countries, or to children in very different cultures, who may have different stressors in their lives and different exercise and eating habits. Remember that the key is whether the children in the study are representative with respect to the relationship between stress and snacking.

Concept 2:
If I change my behavior will it matter?

The Basic Idea

If an observational study finds a relationship between an explanatory variable and an outcome, there is no way to know if the relationship is one of cause and effect. It is quite common for the media to present such results as if there is a causal relationship. The advice will then be that if you change your behavior, or diet, or whatever, you will benefit by the change in outcome. Remember that *a cause and effect conclusion can only be made based on a randomized experiment.*

Test Your Understanding

Question 2.1: Find an example of a randomized experiment in the news that you think may apply to you, now or when you are older. Explain what relationship was found. Do you think that if you changed your behavior based on the explanatory variable (diet, taking aspirin, meditating, etc) you would experience a change in the outcome variable as a result? In general, do you think the differences or changes in the explanatory variable were responsible for a difference in the outcome variable?

Question 2.2: Find an example in the news of an observational study for which the news story or headline is attributing a cause and effect relationship. Explain what relationship was found. Discuss possible confounding variables for the study, or other explanations that may account for the observed relationship. In general, do you think the differences or changes in the explanatory variable were responsible for a difference in the outcome variable?

Question 2.3: Look at the news stories in the Appendix of the book. Identify one story based on an observational study for which the headline is not misleading, and one story based on an observational study for which the headline is misleading. Explain why you reached that conclusion in each case.

Question 2.4: Using the News Stories and Original Sources on the CD accompanying the book, identify two studies, one based on a randomized experiment and one based on an observational study. In each case, explain what behavioral change the study is suggesting, and what the corresponding change is in the outcome. (You don't need to go into detail about the magnitude of the change.) Then describe whether or not you are likely to experience the beneficial outcome if you were to make the behavioral change.

Suggested Answers

Suggested Answer for Question 2.1: This will vary depending on the study. But the basic idea is that for a randomized experiment, you can conclude that if a change is made in the explanatory variable it will cause a change in the response variable, at least on average. (Of course there will be individual differences in response.) Randomization of treatments should minimize the role of confounding variables as an explanation for any differences that are found, unlike in observational studies.

Suggested Answer for Question 2.2: This will vary depending on the study. The basic idea is that there are likely to be confounding variables that differ with the explanatory variable, and that may contribute to differences in the response variable. The confounding variable(s), instead of or in addition to the explanatory variable, may be responsible for the observed relationship between the explanatory and response variables.

Suggested Answer for Question 2.3: The stories conveying results of observational studies are #2, 4, 5, 7, 8, 10, 12, 13, 14, 15, 16, 18, 19 and 20. There are misleading headlines for News Stories 8, 14 and 20 because they are worded to imply that a difference in the explanatory variable *causes* a difference in the response variable. The others indicate that there *may* be a relationship, or that two variables are *linked.* These are legitimate ways to express the results.

Suggested Answer for Question 2.4: See the answer to the previous question for a list of observational studies. Stories 1, 3, 6, and 11 are candidates for the randomized experiment in your response. (Story 9 is based on a meta-analysis and the experiment described in #17 was not done on humans.) You are likely to experience the beneficial change for the randomized experiments, but not for the observational studies. For instance, News Story 1 indicates that if you were to follow a meditation course like the one used in the study, you would be likely to have enhanced immune functioning. (Of course there are individual differences in responses, so probably not every individual will have this experience.) In contrast, using one of the observational study examples, News Story 10 indicates that churchgoers live longer but that does not mean that if *you* start attending church you will live longer than you would have otherwise. There are too many possible confounding variables, and you aren't likely to be changing those as well just by attending church.

Concept 3:
Can I believe the results of that poll?

The Basic Idea

There are many ways in which bias can enter surveys and polls, in addition to the method of selecting the sample, covered in Concept 1. Some of the many factors that can lead to bias include the wording of questions, who asks, the order in which questions are asked and whether they are open or closed questions. Unfortunately, news stories based on surveys and polls rarely give you the details necessary to ascertain problems. When you read the results of surveys and polls, try to find a website that has a more complete description. Many polling organizations provide press releases on their websites for a few days after the results are released, so if you act fast you may be able to get these details.

Test Your Understanding

Question 3.1: It is common for each political candidate in a major election to have his or her own pollster. Why do you think this is the case? If the questions are released and people can see that they are not using biased wording, why do you think it would matter who is conducting the polls?

Question 3.2: A study was done by students in an honors program at the University of California, Davis, to help the Theater and Dance Department increase attendance by students at their events. The study included a survey designed to find out how much students know about the performances given by the Department. One issue of concern was whether students know the location of the theater where these performances are given. Rather than ask, "Do you know where the

Wright Theater is located?" the survey question used was "Where is the Wright Theater located?" From this, it could be determined whether or not respondents knew where it was located. Why is the latter version of the question likely to produce a better estimate of the proportion of students who do know where the theater is located than asking them the direct yes or no version of the question?

Question 3.3: In her book *Tainted Truth* (Simon and Schuster, 1994), Cynthia Crossen noted that even questions about simple comparisons can produce biased results. She described a study presented at a conference, which showed dramatically different results when people were asked "Is tennis more or less exciting than soccer?" compared with "Is soccer more or less exciting than tennis?". Her explanation was that the sport mentioned first was the focus when people answered, and the basis for the comparison. Create a comparison question of this type that you think would illustrate this point, and explain why you think people would respond differently to the two versions. Try out your two versions on two groups of people and report on how they respond.

Question 3.4: Read Section 3.2 in the book (pages 37-41). Find an example of a survey that demonstrates one of the seven pitfalls listed on page 38. Explain which one is demonstrated and how the survey could have been done differently to avoid it.

Suggested Answers

Suggested Answer for Question 3.1: It is to each candidate's advantage to have polls released showing that candidate in the lead. There are many factors other than question wording that can influence the outcome of a poll. For instance, the timing of the poll is crucial. If the incumbent has just given a well-received speech to the nation, that can inflate his or her popularity for a short time. Order of questions is also important. For instance, a poll hoping to show support for a non-incumbent may first ask people how they feel about the state of the economy (if it's poor) or other issues for which the incumbent is not popular. Then, after having led respondents to realize that there are problems with the way things are currently being handled the poll would ask who they plan to vote for in the upcoming election.

Suggested Answer for Question 3.2: Respondents do not like to admit that they are ignorant. Many people would say that they did know where the theater was located, even if they did not. By asking them to specify where it's located, they couldn't pretend to know if they didn't.

Suggested Answer for Question 3.3: This will vary for each person, but the idea is that whatever is mentioned first should invoke a different mental scenario than what is mentioned second. An example might be "Is staying home and watching a movie more or less fun for you than going out to a movie theater?" versus "Is going out to a movie theater more or less fun for you than staying home and watching a movie?". In the first case, respondents would be likely to invoke a scenario in which staying home and watching a movie was fun, and use that as the basis for comparison. They would not invoke the full scenario of going out to the movie theater, but would only think about it in comparison to the advantages of staying home, bringing to mind negatives like paying more for popcorn and having to drive to the movies.

Suggested Answer for Question 3.4: This will vary for each person.

Concept 4:
Can I be unusual and still be normal?

The Basic Idea

In everyday language "normal" and "average" are often confused. For instance, weather reporters talk about the normal rainfall for a given month of the year, or normal high temperature for a given date. What they really mean is the *average* rainfall or high temperature. (Actually, they use a more complicated method than a simple average, but the result is similar.) In the other direction, we talk about whether someone is of "average intelligence" or gets "average grades." We don't mean that they are *exactly* at the average; we mean that they are in a range that includes the majority of people. This wouldn't be such a problem if people understood natural variability, and that for almost all things we measure, there is a range that would be considered normal, but only one single number that is the average.

Test Your Understanding

Question 4.1: Pay attention to the speech of those around (including radio and TV) you until you hear someone use either the word "average" or the word "normal" in the sense described above. Explain the context in which the word was used, and whether the other word (average or normal) would have been more appropriate in that context.

Question 4.2: Explain what is wrong with each of the following two quotes, both of which you may encounter in your daily life. From a friend: "I tend to have a lower body temperature than normal – it's closer to 98 degrees than 98.6 degrees." From a teacher: "I've taught this course many times, and as a class, your results on the midterm were about average for students in this course."

Question 4.3: Refer to Original Source 11 on the CD accompanying the book, "Driving impairment due to sleepiness is exacerbated by low alcohol intake." Two of the treatments used in the experiment involved "normal night sleep." What is meant by "normal" in this context? Is it a single amount, or a range? Is "normal" the appropriate word to use? (Hint: If you search the article for the words "normal" and "mean" - as a synonym for "average" - you will find the appropriate information.)

Question 4.4: Refer to News Story 18 in the Appendix of the book, "Heavier babies become smarter adults, study shows." In the first few sentences the word "normal" is used twice. Is it used appropriately, or would "average" be more appropriate? Explain.

Suggested Answers

Suggested Answer for Question 4.1: This will vary for each person.

Suggested Answer for Question 4.2: In the first case, the friend is not aware that there is natural variability in *normal* body temperature across individuals, and is implying that the often quoted *average* body temperature of 98.6 is the only one that is "normal." In the second quote, the teacher really means that the grades fall within the *range* of grades *normally* seen on the midterm in the course.

Suggested Answer for Question 4.3: The word "normal" is appropriate. A range is provided, 450 to 540 minutes, with a "mean" of 503 minutes.

Suggested Answer for Question 4.4: The word "normal" is appropriate because a range is discussed. The article even provides a specific definition – babies weighing more than 5.5 pounds are "considered to be normal."

Concept 5:
Relationships are not always what they appear to be!

The Basic Idea

Chapter 11 is devoted to explaining various ways in which relationships can come about, in addition to a direct cause and effect. The most common of these ways is that there is a third variable affecting both the explanatory and response variables. A special case is that the third variable is time, and that both the explanatory and response variables are changing over time. The section on Concept 2 already asked you to explore the role of confounding variables in observational studies. This Concept includes many additional ways in which false relationships may turn up.

Test Your Understanding

Question 5.1: Explain in your own words what a confounding variable is. Give an example of where an apparent causal relationship between the explanatory and response variable is probably being influenced by a confounding variable.

Question 5.2: Read Thought Questions 2 and 4 in Chapter 11. Construct an equally absurd example of two variables that would be related and explain why they would appear to be related.

Question 5.3: Explain what type of relationship "correlation" measures. Then explain how and why correlation can be heavily influenced by outliers.

Question 5.4: Give an example of a relationship you have heard about from a source other than an example in the book (such as the relationship between drinking red wine and lower chance of heart disease). For your example, explain which one or more of the "reasons for relationships" on page 207 of the book might explain the relationship.

Suggested Answers

Suggested Answer for Question 5.1: A confounding variable is in addition to the explanatory and response variable. It's likely to differ for different explanatory variable groups, and it's likely to affect the response variable. For instance, in a study where the explanatory variable is whether someone smokes and the response variable is something else related to health, like blood pressure or heart disease, a possible confounding variable could be amount of exercise people get.

Suggested Answer for Question 5.2: This will vary for each person but an easy way to do this is to think of cities as the units, and think of two variables that are both related to the size of a city. For instance, the two variables could be number of hotels and number of babies born during the year. They would be highly related because larger cities would have many of both and smaller cities would have fewer of both.

Suggested Answer for Question 5.3: Correlation measures how closely points in a scatter plot fall to a straight line. Adding an outlier that is far away in both the horizontal and vertical direction will "pull" the line toward it, making it look like the points fall closer to the line, thus increasing the correlation.

Suggested Answer for Question 5.4: This will vary for each person.

Concept 6:
Combining Groups: Beware of Simpson's Paradox!

The Basic Idea

Combining groups or situations when studying a relationship between two variables can completely change the apparent nature of the relationship. A special case (Simpson's Paradox) is when the change is so extreme that the direction of the relationship actually reverses when the groups are combined. Example 7 in Section 12.4 of the book illustrates this extreme case. When you read about a relationship of importance to you, think about whether the people or objects studied might have included different groups, and if so, find out if the same relationship held for all of them.

Test Your Understanding

Questions 6.1 to 6.3 each give an example of a situation where groups have been combined in a way that may cause the results to be misleading. In each case, what groups do you think should have been kept separate? Do you think the relationship in each group would have been in the same direction as the combined groups, the opposite direction, or would the relationship disappear?

Question 6.1: A higher proportion of drivers of red cars were stopped for traffic violations than drivers of white cars.

Question 6.2: A study found that people in large cities are more likely to have their income tax returns audited than people in small towns.

Question 6.3: A study asked college students if they had ever used condoms and if they had ever had a sexually transmitted disease.

Surprisingly, those who used condoms were more likely to have had a disease than those who didn't use condoms.

Question 6.4: Find in the news or invent your own example of a situation where you think combining groups would produce misleading results.

Suggested Answers

Suggested Answer for Question 6.1: There are several plausible answers. One possibility is different age groups. Young drivers may prefer red cars to white cars, and may also be stopped more often. If you compared drivers of red cars versus white cars in each of several age groups, you may find that they are stopped about equally often within each age group. Another possibility is type of car. Perhaps more sports cars are red and more family cars are white. Within type of car, it may be that the same proportion of drivers of each color of car is stopped.

Suggested Answer for Question 6.2: It's probably because very high income earners are more likely to live in large cities than in small towns. The grouping variable is income level. Within each income level, there may be no difference in audit rates.

Suggested Answer for Question 6.3: The logical grouping variable is level of sexual activity. Grouping students into those who are frequently sexually active and those who are not, the relationship would likely be reversed in each group. Within each group, those who use condoms should be less likely to have had a disease. Combining groups is deceptive, because the students who are sexually active are more likely to use condoms and more likely to have had a disease.

Suggested Answer for Question 6.4: This will vary for each person.

Concept 7:
Wow! Was that coincidence really amazing?

The Basic Idea

When a coincidence occurs we tend to focus on the fact that the specific sequence of events had very low probability of occurring. But it is not at all surprising that extraordinary events happen to each of us once in awhile. And it's not at all surprising that events that seem almost impossible, for instance, with probabilities as low as 1/1,000,000, happen to someone, somewhere once in awhile.

Test Your Understanding

Question 7.1: Find a story of a startling coincidence in the news or on the Internet. (One way to do this is to type "amazing coincidence" into a search engine. You will get plenty of material.) Evaluate roughly how likely you think it is that the *specific* sequence of events would happen to the specific person to whom they happened. Then, evaluate roughly how likely *something* similar to that sequence of events would be to happen to the specific person. Finally, evaluate how likely you think *something* similar to that sequence of events would be to happen to *someone, somewhere, someday.*

Question 7.2: Explain why the last sentence in the explanation of "The Basic Idea" above is true.

Question 7.3: If a striking coincidence were to happen to you, would you be inclined to think it was completely explainable using probability or would you think something more mysterious was going on? What questions might you ask yourself to help make this assessment? (Note that this is a philosophical question with no correct answer!)

Question 7.4: Every day, we each have hundreds of experiences, we may encounter many people, and we receive lots of information. Each of these experiences has very low probability of occurring exactly as they do. For instance, if I'm driving and find myself following a white car with the numbers 385 as part of the license plate, then a red car with license numbers 099, it's unlikely that this specific sequence of events would occur again. Yet we don't consider these experiences to be surprising. Why not? What makes a startling coincidence different from ordinary experience?

Suggested Answers

Suggested Answer for Question 7.1: This will differ based on the story, but by the time you make the final assessment (something similar to someone, someday, somewhere) the probability may be quite high.

Suggested Answer for Question 7.2: See the top of page 338 in the book for an explanation.

Suggested Answer for Question 7.3: This depends on your philosophy of life. Some people think that there are "synchronicities" – a term coined by Carl Jung – that indicate there is a higher order to the universe. The best you can do is to try to assess whether more striking coincidences seem to happen to you than could reasonably be expected by chance, but that's a very difficult assessment to make. This is, of course, why we still don't have answers to mysterious questions like this one!

Suggested Answer for Question 7.4: Sequences of events seem surprising and get labeled as coincidences when they have personal meaning for us, or when they seem to involve many unlikely connections.

Concept 8:
Don't get your probabilities backwards!

if P>Q ≠ Q>P

The Basic Idea

The name given in the book for this idea is "confusion of the inverse;" see Section 18.4, pages 340-342. The idea is that we are interested in the probability of a particular event *given* knowledge of another event. We confuse this with the probability of the second event, *given* knowledge of the first event. The example in the book is that many physicians think that the probability that you have a disease *given* that your test is positive is the same as the probability that you will have a positive test, *given* that you have the disease. These can be quite different.

Test Your Understanding

Question 8.1: Refer to Original Source 5 on the CD accompanying the book, "Distractions in Everyday Driving." Find Table 1, in which results of a 1999-2000 Pennsylvania study are shown. Notice that of drivers involved in a crash for which there was a distraction, only 5.2% were using or dialing a cell phone, whereas 21.9% were distracted by an "outside object, person or event." Would it be appropriate to conclude that the probability of having an accident given that the driver is using a cell phone is lower than the probability of having an accident given that the driver is observing an outside object, person or event? Explain.

Question 8.2: Refer to Question 8.1. Construct a table similar to Table 18.1 in the book, based on a hypothetical 100,000 drivers. Suppose 1,000 of them have an accident and 99,000 don't. Based on the data in Question 8.1, assume that 5.2% of those who have an accident are using a cell phone and the remaining 94.8% aren't. Finally, based on data given

83

in Original Source 5, assume that at any given time 1.5% of all drivers are using a cell phone in the car and 98.5% of drivers are not. After constructing your table of one hundred thousand hypothetical drivers, find the probability that a driver has an accident *given* that they are using a cell phone. Compare it to the probability that a driver has an accident *given* that they are not. Comment on this result.

Question 8.3: Explain how the lottery slogan "You have to play to win" may be an example of misleading people through confusion of the inverse.

Question 8.4: Refer to Figure 17.1 on page 329 of the book, and the accompanying description. It shows that if a weather forecaster says it is definitely going to rain, the probability that it actually will rain is about .91. Would you prefer to know that, or would you prefer to know the answer to this question: Given that it actually does rain, how likely is it that the weather forecaster said it would rain? Now apply this same question to the situation described for physicians. Would you rather be able to answer "Given that my physician definitely says I have the disease, what is the probability that I have it?" or "Given that I have the disease, what is the probability that my physician says I have it?" Do you think those two probabilities are similar? Explain.

Suggested Answers

Suggested Answer for Question 8.1: No. This is an example of confusion of the inverse. *Given* that a driver had an accident based on a distraction, the probability is about .052 (5.2%) that they were using a cell phone and about .219 (21.9%) that they were distracted by something outside. That doesn't tell us anything about the probability of having an accident *given* that the driver is using a cell phone.

Suggested Answer for Question 8.2: The table would be as follows:

	Cell phone use	No cell phone use	Total
Accident	52	948	1,000
No Accident	1,448	97,552	99,000
Total	1,500	98,500	100,000

Therefore, the probability of having an accident given cell phone use is 52/1500 or .0347. The probability of having an accident given no cell phone use is much lower, at 948/98,500 or .0096. The relative risk is about 3.6. This is a good example of confusion of the inverse. The probability of having an accident *given* cell phone use is quite high, compared to the probability of an accident *given* no cell phone use, even though the probability of cell phone use *given* an accident is lower than the probability of no cell phone use *given* an accident. (Remember that these numbers are illustrative only and may not represent the actual probabilities. More baseline data would be needed to determine accurate probabilities. Also, cell phone usage has increased since the study was done.)

Suggested Answer to Question 8.3: Although what the slogan is really saying is that the probability that you played, *given* that you won, is 1, some people may subconsciously switch it around and think that the probability that you will win, *given* that you play is also close to 1. It would be more accurate to say that "you have to play to win, but you don't have to win if you play!"

Suggested Answer to Question 8.4: For the weather forecasters, the probability in the direction given is the one of interest. But for the physician, you would probably be more interested in knowing the probability that your physician can detect a disease if you have it, than you would be in knowing the probability that you have a disease given that your physician says so. The two probabilities can be very different. Thinking they are similar is confusion of the inverse.

Concept 9:
Knowing when not to trust your intuition.

The Basic Idea

Psychologists have shown that we do not have very good intuition about assessing probabilities in everyday circumstances. It is very easy to be misled unless you understand how you can be manipulated. Think carefully when you have to make a decision based on incomplete information. Make sure you are familiar with the psychological influences on probability discussed in Chapter 17.

Test Your Understanding

Question 9.1: Suppose you are buying a new television and it comes with a one-year warranty. The salesperson explains that you can buy an extended warranty that will cover you for an additional two years. How would you decide whether or not to buy it?

Question 9.2: Do you believe that there is intelligent life in the universe, other than on earth? (We won't discuss whether there is intelligent life on earth!) Answer yes or no. What do you think is the probability that you are correct? Whichever way you answered, now list reasons why you might be wrong. Take this seriously, and think about all of the possible reasons you could be wrong. Question any basic assumptions you may have. Now reassess the probability that you are correct. Did this exercise change your viewpoint at all? Should it have, based on research by psychologists? Discuss.

Question 9.3: When we hear about a relationship found in an observational study, it is sometimes very tempting to conclude that there is a causal relationship. For instance, in Case Study 16 on the CD, a relationship was found between exposure to TV violence as a child and violent behavior later. It is tempting to conclude that watching violent TV causes violent behavior. Which one of the following two psychological influences is more likely to explain this tendency to attribute a causal connection: anchoring or representativeness? Explain.

Question 9.4: Explain how anchoring is used when trying to sell products on television or the Internet.

Suggested Answers

Suggested Answer for Question 9.1: First, don't be misled by detailed scenarios that the sales person might provide for what could go wrong. The cost of an extended warranty is based on the manufacturer's estimate of what proportion of televisions will fail, and what it will cost to fix the ones that fail. They compute an "expected value" or an average cost per unit for repairs, including those that fail and those that don't. Therefore, you are smart to buy the extended warranty only if you think your television is more likely to fail than other ones sold. For instance, if you know it will get much heavier use than the typical television, or you know you plan to move it around often as you change apartments, then your individual probability of needing a repair may be high and it may be a smart purchase.

Suggested Answer for Question 9.2: Read the material in Section 17.4, pages 326-327. According to research, this exercise should have changed your probability of being right somewhat. This is probably because of the availability or representativeness heuristic, depending on the reasons you specified. If you can imagine reasons why the other conclusion might be justified, or readily bring to mind scenarios of other intelligent life, you may move your personal probability somewhat in its direction.

Suggested Answer for Question 9.3: Representativeness could lead us to assign a high probability to a causal connection. On page 324 of the book it's stated that "the representativeness heuristic leads people to assign higher probabilities than are warranted to scenarios that are representative of how we imagine things would happen." In the example of TV violence, it's easy for us to imagine that watching violent television would lead people to be angry, which would lead to violence.

Suggested Answer for Question 9.4: The idea of anchoring is to present you with one price first, but then explain that you will be able to buy the product for a lower price. This is done routinely in commercials on television, where the announcer might say that you could expect to pay $25, $30, or more for this product, but if you act now you can buy it for only $15. On the Internet, often one price will be presented, and then crossed out and a "sale price" will be given instead. The consumer uses the originally stated price as an anchor, and thinks anything less is a bargain. In truth, even the "bargain" price may be inflated.

Concept 10:
What's important may not be significant!

The Basic Idea

If a study fails to find a statistically significant difference or relationship, don't interpret that to mean that there isn't an important difference or relationship in the population. Unless the number of individuals in a study is quite large, even a moderate population relationship may not produce data that achieves statistical significance. If possible, examine a confidence interval to find out if it falls mainly to one side or the other of chance. Be particularly wary of studies based on a small number of individuals.

Test Your Understanding

Question 10.1: Original Source 9 on the CD presents this conclusion (on page 790): "These findings fail to support an overall difference in suicide risk between antidepressant- and placebo-treated depressed subjects in controlled trials." The suicide rate for those taking antidepressants was 0.76%, with 95% confidence interval of 0.49% to 1.03%, whereas the suicide rate for those taking placebos was 0.45%, with 95% confidence interval of 0.01% to 0.89%. Explain what the authors mean by saying that the findings "fail to support an overall difference." Can they conclude that there is no difference in effectiveness of antidepressants and placebos in the population?

Question 10.2: Read the explanation of the "power" of a test, on page 424 of the book. Now explain what power means in the following context. In a presidential election, suppose the incumbent (current president) actually has the support of 60% of the voters in the population. A poll is to be taken, and a one-sided hypothesis test will be done. The null hypothesis is that 50% (or less) of the population support the incumbent and the alternative hypothesis is that more than 50% (a majority) support the incumbent. If a random sample of 100 people is taken, and the usual significance level of .05 is used, the power of the test is only .64. Explain what this means.

Question 10.3: Refer to the previous question. Would it be more desirable to increase or to decrease the power of the test? What could be done to accomplish that?

Question 10.4: Read Case Study 25.2, which begins on page 477. One of the quotes reported on the top of page 478 states that no difference in mortality from breast cancer was found for women aged 40-49 that could be attributed to mammogram screening. After reading the Case Study, discuss that conclusion in the context of the Concept here, "What's important may not be significant."

Suggested Answers

Suggested Answer for Question 10.1: The confidence intervals for the suicide rates overlap substantially, so it isn't clear whether there is an overall difference in the population. But, that's not the same thing as concluding that there is definitely no difference. A larger study would produce more narrow confidence intervals and may result in intervals that don't overlap.

Suggested Answer for Question 10.2: The power of the test is the probability that the null hypothesis will be rejected. In this case, the probability is only .64 that the poll will actually provide convincing evidence that the incumbent has the support of a majority of voters, even though 60% of the population supports her or him.

Suggested Answer for Question 10.3: It would be more desirable to increase the power of the test, thus increasing the probability of making a correct decision. Assuming the standard significance level of .05 is still used, the only way to increase the power is to take a larger random sample for the poll.

Suggested Answer for Question 10.4: What the authors really meant to say was that they didn't find a *statistically significant* difference in mortality based on mammogram screening. They did in fact find a difference for the sample data, and it should not have been reported that they found no difference. Readers would interpret that to mean that they concluded that there is no difference in the population, rather then to mean that they simply didn't have enough data to know.

Concept 11:
What's significant may not be important!

The Basic Idea

Statistical significance is not the same thing as practical significance or importance. Almost any null hypothesis can be rejected if the sample size is large enough. Be careful when you read the results of a large study. Try to find a correlation, relative risk, difference or other summary that will give you some idea of the magnitude of the effect. It's even better if you can find a confidence interval to accompany the summary.

Test Your Understanding

Question 11.1: Explain what it means to say that a study found a statistically significant relationship. To help you with your explanation, use Case Study 3 on the CD, which found that cholesterol dropped significantly in a group that ate a special diet for a month. In other words, the drop in cholesterol was statistically significant. Explain what that means.

Question 11.2: Explain what is meant by a *p*-value. To help you with your explanation, use Case Study 3 on the CD. One of the findings was that there was a drop of 8% in cholesterol for those who were on the "control" diet – a very low-fat diet. The accompanying *p*-value was reported to be 0.002. Explain what the *p*-value is.

Question 11.3: Refer to the previous question. Which of the reported numbers, 8% or 0.002, should be the focus if you are trying to decide whether the result is statistically significant? Explain.

Question 11.4: Refer to the previous two questions. Which of the reported numbers, 8% or 0.002, should be the focus if you are trying to decide whether the result is of practical importance? Explain.

Suggested Answers

Suggested Answer for Question 11.1: It means that *if* the diet would really have no effect on cholesterol in the population, then we would have been very unlikely to see such a strong effect in the sample.

Suggested Answer for Question 11.2: The *p*-value is the probability of observing a drop of 8% or more in a sample of the size used, *if* there would be no drop on average for the population if they were to go on this diet.

Suggested Answer for Question 11.3: The important number in this case is the *p*-value of 0.002. Because it's less than 0.05, the reported drop of 8% is statistically significant.

Suggested Answer for Question 11.4: The important number in this case is the drop of 8%. Knowing whether that's of practical importance requires knowledge of whether a drop in cholesterol of that magnitude matters for any health outcomes.

Concept 12:
Why does one out of 20 foods seem to cause cancer?

The Basic Idea

Using a level of significance of .05 means that for all tests in which the null hypothesis is true, we will decide to reject the null 5% (.05) of the time by chance. (This is not quite true if the tests use data from the same individuals and are thus not independent, but it will be close.) That means that in a large study for which many relationships are investigated, about 5% (1 in 20) of all of the ones that don't exist will appear as if they do exist. For instance, if many foods are tested for their relationship to a disease, and none are related, about one in 20 will appear to be related. There are sophisticated methods for dealing with this problem, called "multiple comparisons." But when you read reports of studies in the news you would not generally be told if they were used. Therefore, use caution when interpreting results that were clearly part of a larger study. News articles may mention that other things were tested, but will generally focus on the results that were statistically significant and ignore the rest.

Test Your Understanding

Question 12.1: Is this problem equivalent to saying that one in 20 statistically significant results are erroneous, and are just due to chance? Explain.

Question 12.2: Find an example of a study in which it is clear that multiple hypotheses were tested. Comment on whether the news article mentioned that as a problem. Discuss the extent to which you think this issue affected the conclusions made in the news story.

Question 12.3: Would at least 20 hypotheses have to be tested in a study in order for the issue raised in Concept 12 to be a problem? Explain.

Question 12.4: What do you think reporters should do to help readers of news reports with this issue? Reading the section on "Multiple Hypothesis Testing" beginning on page 498 of the book may help you answer this question.

Suggested Answers

Suggested Answer for Question 12.1: No, in fact that's a good example of confusion of the inverse (Concept 8). We are concerned with the probability that a study will find statistical significance *if* chance alone is at work. This question asks about the probability that chance alone is at work, *if* the results are statistically significant. There is no way to determine that.

Suggested Answer for Question 12.2: This will vary for each person.

Suggested Answer for Question 12.3: No. Anytime more than one hypothesis is tested it could be a problem. However, the more hypotheses tested, the more likely it is that this will be a problem.

Suggested Answer for Question 12.4: All of the hypotheses included in the study should at least be mentioned, so that educated readers can assess whether or not there is likely to be a problem.